Wild Horses
and
Sacred Cows

Wild Horses and *Sacred Cows*

by Richard Symanski
Foreword by Edward Abbey

Northland Press • Flagstaff, Arizona

The publisher and author would like to acknowledge their
appreciation to *Science* magazine for permission to include
material from Garret Hardin's article, "The Tragedy of the Commons," which appeared in vol. 162, pages 1243–1248, 13 December 1968 (© 1968 by AAAS); to Hope Ryden, author of
America's Last Wild Horses (revised edition, New York: Dutton,
1978) for information on the trial of Cook, Barber, et al.; to Tina
Nappe, for permission to include the Gus Bundy photographs
on pages ii and 7; to the Bureau of Land Management, for permission to reproduce the photographs on pages 29, 49, 84, 164;
to *The Daily Rocket-Miner* of Rock Springs, Wyoming, for permission to reproduce the photograph on p. xvi; and to Skylar
Hansen, author and photographer of *Roaming Free: Wild Horses
of the American West* (Flagstaff, AZ: Northland Press, 1982), for
permission to reproduce the photographs on pages 42, 122, 182,
206. A final, special acknowledgement is due Edward Abbey for
his permission to expropriate the title of his foreword.

*For Nancy,
my arresting biologist*

CONTENTS

FOREWORD

Wild horses! What words, what images, what memories, best evoke the essence of the American West? These are some of the ones that come first to my mind: the smell of sage crushed in the hand. The fragrance of burning juniper. A mountain lion crouched on a canyon ledge. The word "canyon" itself. One black vulture soaring in lazy circles above the Painted Desert. Red mountains like mangled iron rising above dunes of sand. Stone ruins nestled in an alcove of a canyon wall. The cry of coyote— first one, then a second, then an ululating chorus—when a full moon the color of blood sinks beyond the western skyline. The aroma of burning mesquite. One dust devil spinning across an alkali flat. An abandoned Model-T Ford sunk fender-deep in sand along a back road in the Arizona Strip country. The sound, at night, of something bulky and *fierce* crashing through an alder thicket on the slopes of Two Medicine Mountain in Glacier National Park—the smell of the grizzly. Red and yellow billboards along old U.S. Highway 66 warning the westbound motorist: 200 MILES OF DESERT AHEAD LAST CHANCE FOR WATER. A real Indian on a pinto horse. A genuine old-time ranger with Smokey Bear hat and uniform of forest green, riding a big bay horse and leading a string of pack mules down the switchbacks at the Bright Angel Trail, down and down and down into the mystic depths of the Grand Canyon. The smell and creak of saddle leather. The clank of a turning windmill near a broken-down corral. The smell of horse dung. The smell of horses. And the first sight, at evening, of a file of slick, unbranded, unclaimed, tangle-maned and broom-tailed mustangs coming down the ridge for water, old mare in the lead, a stallion at the rear. Wild ones.

The romance of the wild horse haunts the American West as much today as it ever did in the past. As Richard Symanski explains and documents in his most thorough and comprehensive book, the wild horse or "mustang" is still very much with us, even thriving and multiplying, at least in parts of Nevada, southeastern Oregon, southwestern Wyoming, and the western deserts of Utah. Like the feral burro in Death Valley and Grand Canyon, the ever-growing herds of wild horses are becoming a nuisance, a bother, a destructive element in those areas where they have been allowed to roam unhindered, protected by an Act of Congress in 1971 from their only serious enemies—cowboys and the pet-food industry. Until that year, most wild horses ended their careers inside tin cans on supermarket shelves. Protection of the wild horse has led to what could have been foreseen by any biologist: a rapid increase in the number of wild horses. Like fruit flies, like rabbits, like humans, the wild horse is a sexy animal, a prolix and fecund beast, and given the opportunity, it multiplies with zest, at an exponential rate. These growing numbers in turn impose a constantly heavier burden on the carrying capacity of the range. Horses graze, and like cattle, they also browse when sufficient grass is not available. Thus, they compete not only with cattle but with other herbivores—deer, elk, bighorn sheep, pronghorn antelope.

Hunters and game-and-fish departments in the western states have apparently been slow to perceive that the growing numbers of wild horses will eventually threaten the forage of their own favorite form of wildlife. Not so the operators of the public lands beef industry. For many years, western cattlemen have tolerated a small number of wild horses, because they saw the animals as a fairly reliable source of income (dog food) and as occasional replacements for draft animals and saddle mounts. But as increasing numbers of wild horses compete directly with their beef cows, the ranchers have become agitated. They are demanding in stronger and louder language that the federal government control the horses' numbers. At the same time, public sentiment—or horse-loving sentimentality—resists attempts to return the mustang to its former status as raw material for a commercial product. This is the core of the controversy with which this book is concerned.

What are my own feelings on the matter? Like the author of this book, I think that the health of the land and the well-being of our native wildlife come first. If, as seems obvious, the wild horses are a menace to both, then we should adopt whatever means are necessary to reduce and limit their numbers. I would suggest confining the remainder within a few large desert mountain reserves that are also well-stocked with mountain lions, jaguars, and wolves.

But I have no sympathy whatsoever for the public-lands beef industry. The eleven western states produce about three per cent (3%!) of the total American beef production. There are more beef cattle in Vermont than in Montana. There are more beef cattle in Florida and Georgia than in all of the western states combined. We must overcome our juvenile fascination with the mythology of the cowboy and rancher and recognize the industry for what it actually is: one that has been getting a free ride on public property—our property—for more than a century. There is no corner of the American West, aside from the national parks, that is not infested with domestic cattle. There are more cattle at large today than there were in the 1880s, the legendary glory days of the business. Everywhere you may go in the mountains, canyons, forests, or deserts, you are nearly certain to encounter hordes of these shambling, ugly, stupid, dung-besmeared, sore-eyed, disease-ridden, fly-spreading, bawling, drooling, heavy-footed, and voracious brutes. The damage caused by wild horses to the public rangelands is so small, so miniscule, so recent, in comparison to the awesome and prolonged devastation of the beef-cattle industry, that the two are not commensurable.

We need at least a few herds of wild horses here and there, if only to preserve a beautiful tradition, but we have no need whatsoever for private cattle on public lands. It is one of the many merits of Mr. Symanski's fine book that he makes us aware not merely of the complicated problems resulting from wild horse protection but also of the far bigger, far more important issue that looms beyond: To whom do the federal lands belong? For whose benefit should they actually be managed? And are human needs the only needs worthy of consideration?

Edward Abbey
Oracle, Arizona

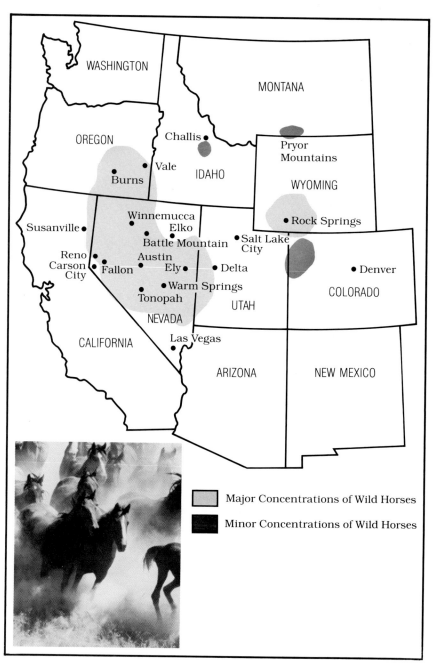

Wild Horses on Western Rangelands.

PREFACE

I've long had a love affair with Nevada. Not with the hot lights and the fast lanes of Las Vegas or Reno, but rather with the shadowless expanses of pungent olive-green sage and shimmering tan and white that turn mauve and mysterious with nightfall; with untrodden mountains mottled with huddled junipers, twisted pinyons, and wildflowers with names like skullcap and bladderpod, mule's ear and monkshood; with navelike canyons inhabited by loggerhead shrikes, red-shafted flickers, chukars, mule deer, antelope, bobcats, kangaroo rats, vagrant shrews, diamondback rattlers. And mustangs.

Any excuse is excuse enough to find myself in Nevada, or for that matter, the desert West. When I looked for a new reason in the spring of 1982, my wife, an evolutionary biologist willing to feed my passionate predilections, suggested that I do a story on the West's mustangs. "But I know nothing about the horses," I replied. "You know they're in Nevada in abundance," she exclaimed, laughing. And, she might well have added, "You can borrow or steal or lean on my knowledge of biology at will." So, without further ado, I launched on a venture of unperceived proportions, with a bias no greater than that of a desire to preserve the remote desert heartland of Nevada.

Although Nevada has more than half of the nation's mustangs, there are also significant free-roaming horse populations in Wyoming, Oregon, and California, with lesser numbers in Colorado, Utah, Montana, and Idaho. Though my heart told me to devote my major effort to Nevada, if I was to be true to my academic background I knew that I couldn't slight lesser places and smaller concentrations of mustangs; I don't believe I did. By September 1984, when I decided that I had to bring closure on what I had learned about the management of the West's wild horses, I had travelled more than fifteen thousand miles in eight Western states. I had watched numerous roundups of mustangs and traipsed in several mountain ranges in pursuit of them. I had talked to more than two hundred people: cattlemen, sheepmen, youthful buckaroos; old-time mustangers and modern-day outlaws; concerned conservationists and wild-horse activists; and dozens of cowboys and desk-bound bureaucrats in virtually every district Bureau of Land Management office where there are mustang populations. With rare exception, those I approached and invited to talk about wild horses were generous with their time and patient with my prying.

I want to thank specific individuals without whom this book would be less than it is: Jack Steinbrech in the BLM office in Rock Springs, Wyoming, who took me on my first wild-horse roundup and began the slow process of banishing my ignorance and innocence; Ron Crowe, who freely gave me a swooping hawk's view of helicopter mustanging; Dawn Lappin of Wild Horse Organized Assistance (WHOA), a tireless champion of the mustang cause, who spent long afternoons in restaurants and in her office throwing out more information and insight into Nevada's wild horses and BLM management of public lands than I was able to digest; Milt Frei, Nevada's BLM wild horse honcho, who provided valuable maps, BLM reports and memos, and enough maverick opinions and interpretations of BLM successes and mistakes and policies to keep my adrenaline flowing for hours; Andy Anderson, range and wild horse specialist in the Carson City BLM office who always had time for my pestering questions, and who proved to be a rich repository of sage observations on the wild horses of western and central Nevada; George High, the Nevada BLM's chief field investigator for "wild horse irregularities," who supplied eye-opening insights into illegal mustanging and the lore of cowboy Nevada; Joe, Sue, and Helen Fallini, who graciously put me up on their Twin Springs Ranch and took time out in the middle of their dawn-to-dusk summer roundup to make sure that I thoroughly understood their problems and their frustrations with mustangs; John McClain, who showed me films and introduced me to ranchers and his work on environmental impact; Jerry Wilcox of the Vale, Oregon, BLM office, without whom my knowledge of the history of mustanging in Oregon would be sketchy in the extreme; and Tina Nappe, who educated me on the Sierra Club's position on mustangs, told me what to look for in all those "little girl books" on horses, and saw to it that I was kept up-to-date on Nevada wild horse happenings right down to the time the book went to press.

My fieldwork was buttressed by extensive library work at Nevada BLM headquarters in Reno and at several city, county, and university libraries throughout the West. To supplement and cross-check anecdotal material and oral histories, I've relied on scores of articles published in popular and scientific journals. Two books that proved particularly useful on the history of mustanging were Anthony Amaral, *Mustang: Life and Legend of Nevada's Wild Horses* (1977) and J. Frank Dobie, *The Mustangs* (1934). Hope Ryden, *America's Last Wild Horses* (1978), was valuable for an appreciation of the protectionist position as well as for court testimony by Julian Goicoechea, Art Cook, and Ted Barber. Heather Smith Thomas, *The Wild Horse Controversy* (1979), was indispensable for background on Wild Horse Annie and on the battle between the BLM and the American Horse Protection Association over Idaho's mustangs.

I want to thank Susan McDonald, my editor, who took a personal interest in the topic and in the execution of the final product, and whose keen eye drew my attention to the errant phrase or paragraph; and Gunther Stuhlmann, my agent, who redirected my efforts on an early draft.

First and last on the list of those to whom I'm indebted is Nancy Burley. She was unfailingly there when I needed help, when I forgot the simple dictum that "less is more," and when I fell into the spidery pit of speciousness. Nancy's Aristotelian mind can be found on every other page.

CHAPTER 1

ROUNDUPS

Inside the hanger of Pinebelt Helicopter, a broken tail from a Bell 47 leans against a wall. An enraged wild horse is crayoned in black on the red and white remains. Above the horse is the figure "$25,000," the replacement cost for a tail kit. Ron Crowe, Nevada's number one whirlybird mustanger, had lurched too low to the ground on a Bureau of Land Management roundup and hit an unyielding rock.

It was a Sunday morning in late July and Ron had just agreed to take me along on a privately contracted roundup on the 102 Ranch in the Virginia Hills, twenty miles north of Virginia City. After strapping on jerricans, Ron ran through instrument checks, revved the engine, and made some notes in his log book. Within minutes we were rising above the Cessnas and the gliders that litter Minden Municipal Airport.

By Ron's own account, there is nothing particularly unusual about his job. He talks about the five thousand wild horses he has helped the BLM round up in Nevada and California the way a tired truck driver describes an uneventful cross-country run. He plays down his good reputation, the fact that when the BLM or a rancher needs a savvy, dependable pilot they know who to call. For the right price, Ron will put one of his helicopters down just about anywhere. One weekend he'll land on the shores of a remote Sierra lake; the next weekend he'll fly the Hiltons to their spacious ranch near Yerington or take them to a little-known river to fish for brown trout. In the spring and winter

Ron makes frequent trips into the Sierras on power line patrols, to measure snow accumulation, the buildup in streams. It's all in a day's work and, as often as not, he looks forward to getting home where he can be with his family, have a beer, and lose himself in a baseball or football game.

For several minutes as we rose skyward we seemed motionless, as if suspended from strings held by a nervous puppeteer. We were doing almost sixty miles an hour as we climbed toward 7,000 feet. Everything was beginning to miniaturize: unsightly mine tailings, a rambling pickup, a colony of trailers, and long circling roads and trails coming from nowhere and disappearing beyond sight. The lumpy tan-and-brown desert floor suggested desolation, alienation, chicken-pox scarring.

Soon, we were over the Pinenut Mountains: yawing, roaring, pitching, shooting to and fro into the unfiltered blue, now and again seemingly on a disaster course with a saw-toothed wall of rock or a dark tessellated line of junipers and pinyons. Ron leaned toward me and shouted, "Look for horses." I looked fore, left, right, down. I saw nothing. Cool air whistled through the open cockpit. Ron pushed on a peddle and moved his right arm; we charged toward a forbidding ridge, then hopped up and over with mere yards to spare. Then, suddenly, it felt like someone had thrown us into an elevator shaft. We dropped fifty, a hundred feet—more, in what seemed like seconds. Ron didn't crack a smile. Seconds later we were zooming straight for another slab of forbidding mountain. A piercing whump-whump and a fwap-fwap-fwap rebounded off the jagged granite. Again I turned to look at Ron. He was peering down and to his left. Suddenly he pitched the extraordinary whirring machine steeply. I grabbed for the belt around my waist, but lost interest in my own welfare when below us, appearing out of nowhere, were several hefty rumps on a thin trail that meandered through a green canyon. A procession of wild horses was leisurely strolling toward the mountaintop—four bays, a couple of sorrels, a small bluish one, and, bringing up the rear, a brown stud. Ron returned for a second pass a football field's length behind them. The horses hardly seemed to notice; several gave no more than a shake of a scraggly tail. As we headed east, Ron shouted out the hard-won wisdom that if you fly directly over mustangs a dozen or so hands above their heads, they might tell you to go fly a kite when you return in a month or so to chase them toward a trap. They can be pushed and cajoled with the unfamiliar fwap-fwap so long as you don't give them a chance to understand that there's not much fighting muscle behind the noise.

Within minutes we saw horses everywhere. Here a band of eight heading into a thick jungle of juniper, there six forming a convivial circle in a rocky clearing, in the near distance two adults and a colt lazily crossing a gulley. For a brief moment I let my imagination get the best of me: I imagined that the threesome was on its way to brunch at the whinnying Mustang Restaurant. Letting go of the reverie, I saw roans, sorrels, greys, blacks, a buckskin or two, a prize paint, a preponderance of sunlit bays. Bays, I would discover, seem to have won the genetic war among Nevada's wild ones.

As we flew over a small wood shack, Ron said that a biology graduate student from the University of Minnesota was living there. The student, Greg McMahon, flew frequently with Ron in the Pinenuts. McMahon was putting large red and yellow collars on horses, releasing them, and then returning later to see how many he could find. One in four of the horses was equipped with a radio transmitter; some had been shot from the helicopter with a blast of gaudy paint. The student and his professor wanted to find out all they could about the horses' daily and seasonal traveling habits: where they eat, where they mate, where they play. But the grand purpose of the $270,000 research project sponsored by the National Science Foundation was to determine how accurate BLM cowboys are at counting wild horses. Greg McMahon was finding that in rugged mountains like the Pinenuts with lots of pinyon, juniper, and big sage, the BLM undercounts the number of horses by as much as fifty percent. In lighter cover, as in the Pah Rah Range north of Reno, BLM tallies are below actual figures by fifteen to twenty percent. With no tree cover, the error drops to five percent or less. McMahon has discovered that the size of a wild horse band affects what an airborne cowboy sees; a band of five or more mustangs will be undercounted by a third. Also, censuses taken in helicopters are better than those done in dwarf-windowed, fixed-wing planes. The little Supercubs fly with their noses up and they don't make the helicopter's sinister whirring noise that usually makes horses move and thereby easier to spot. Finally, McMahon learned that previous experience in counting horses doesn't mean much, and double-counting of the horses is rare. The findings were good news to those in the BLM, which is constantly charged with overcounting the horses by wild horse lovers who have already made up their minds that there are fewer horses in the West than the BLM claims.

It was time to find the horse trap on the 102 Ranch and get down to business. Ron put us on course for Carson Valley and soon we had a lofty, wide-angle view of civilizing artifacts: a hundred or so cows chewing their cuds or eating the grey-green carpet along the Carson River, sprawling five-bedroom homes with swimming pools and riding horses nearby, a Euclidian nightmare of trailers, and junk-filled yards reminiscent of backroads of Appalachia. We crossed Highway 395, which connects Reno and Carson City. There, the inhabitated flats give way to nobs and knolls and ravines that form the lower reaches of the Virginia Hills.

I came to attention when Ron pointed to a dry stream bed and a deep, cuplike canyon. Its downslope opening was closed off by a heap of corral posts that resembled pickup sticks. He said that it had once been an Indian trap for catching mustangs. Before I could get a good second look at the trap site, we had topped an archipelago of islanded summits. We were now in the Virginia Hills. We were in Storey County. We were in Wild Horse Annie country.

Wild Horse Annie's real name was Velma Johnston. She was born in Reno in 1912. Her father freighted goods between California and Nevada, and occasionally hired out as a mustanger—he rounded up wild horses and sold them to the highest bidder. Velma Johnston and her husband owned the Double Lazy Heart Ranch in Storey County, which, in the summer and on weekends, was run as a dude ranch for children. One day in 1950, while on her way to her secretarial job in Reno, Mrs. Johnston found herself poking along behind a large cattle truck full of wounded horses. Disturbed by what she saw, she followed the truck to a rendering plant and there, to her dismay, saw that some of the horses had been blasted with buckshot. A colt inside the truck had been trampled to death. Others looked miserable. Unable to make sense of this cruelty, Mrs. Johnston went to the BLM office in Reno to report what she had seen. She received little consolation from the BLM, and, in fact, discovered that this watchdog arm of the Department of the Interior encouraged catching of wild horses. Mustangs, she was informed, were pests; they consumed valuable grass and drank vast quantities of water that belonged to cattle and sheep and wildlife. The Grazing Service, forerunner of the BLM, had set policy on the matter in the mid-1930s when its first director said that he wanted "the removal of wild horses from public lands." From time to time, the BLM and the U.S. Forest Service had "close-outs" on the animals. Ranchers were given sixty days to round up their own free-roaming horses. The rest were to be shot. As for the inhumane treatment of the mustangs, well . . . nobody liked that. Nor could anyone with a heart like transportation regulations known as "killer rates." Trucks and trains full of horses headed for slaughter were exempted from the need to stop and feed the animals.

Mrs. Johnston didn't like what she heard, and most of all she didn't like the cruelty that she had witnessed firsthand. She took her case to the newspapers, and later she and her husband began a relentless campaign on behalf of the horses. Her first real victory came in 1952 when, with the help of almost one hundred fifty petition signatures, she convinced the Storey County Board of Commissioners not to issue a license for a proposed roundup. (Counties throughout the West have long had the power to authorize wild horse gatherings.) Her victory was a sweet one, for the very men who had applied for the Storey County license owned the rendering plant where she had seen her first example of inhumane mustang treatment.

In June of 1952, Storey County, pushed by the indefatigable Mrs. Johnston, became the first county in the nation to ban the spotting or chasing of wild horses by airplane. But this was not enough for the woman soon to become famous; she wanted a similar law to apply everywhere in the state. Soon, Mrs. Johnston was writing magazine articles and sending letters to newspapers, humane organizations, prominent citizens, riding groups, friends, anyone who might give her a hand. She petitioned state legislators and urged them to sponsor a bill that would make it illegal to chase horses or burros with airplanes or motorized vehicles. Finally, in 1955, state legislators passed a bill similar to the Storey County code. Though the first state law of its kind, it was, in fact, a hollow victory for the woman who by now had acquired the name of Wild Horse Annie. An amendment had been attached to the bill which exempted the state's public lands. Eighty-six percent of Nevada belongs to the public.

By this time, Wild Horse Annie was defining her mission in national terms. She wanted federal legislation to protect wild horses, for though Nevada had more than half of the West's total, there were also significant populations in California, Oregon, and Wyoming. In 1957, her cause received a major boost when California's *Sacramento Bee* published a series of front page articles on mustanging and the pet-food fate of many of the horses. From young and old, from horse lovers and those who had never been in a saddle, Wild Horse Annie received piles of letters asking what could be done. To one and all, she suggested that they contact their representatives in Washington. By 1958, the letters were having an impact. That year, Nevada's U.S. Congressman, Walter Baring—despite pressure from ranchers—introduced a wild horse bill in the House of Representatives.

Thwarted by adjournment, Baring tried again in 1959. This time he was successful, thanks in good part to Wild Horse Annie's trenchant testimony.

Wild Horse Annie argued that in the early 1900s there were around two million wild horses in the West. (Half this many was probably closer to the truth.) Because of the demands of chicken- and pet-food factories in the 1920s, their numbers had declined to one hundred fifty thousand by the mid-1930s. After World War II, professional mustangers increasingly used airplanes to chase horses, and in little more than ten years the population fell to forty thousand. Wild Horse Annie and others were certain that the West's mustangs were in imminent danger of extinction.

Wild Horse Annie wanted a federal law because she believed the use of airplanes encouraged inhumane practices. Airplanes made it easy to maim horses with sawed-off shotguns or run them to exhaustion. Colts who couldn't keep up with a fleeing herd were left behind to die; to a profit-minded pilot they weren't worth the bother or the extra gas. Other methods of gathering horses could be equally cruel. Occasionally, when mustangers on horseback brought in a string of mustangs, they would revive the Paiute custom of nose hobbling. The horses' nostrils were slit with a pocket knife, then baling wire was put through the slits and twisted tight to impair breathing and make escape difficult. Or they might tie heavy bolts in the forelocks to bang against a horse's head when it tried to make a run for freedom.

The U.S. congressional committee was shown photographs strikingly reminiscent of scenes in *The Misfits*, the 1961 Clark Gable and Marilyn Monroe movie in which mustangers ran down wild horses with jeeps and airplanes and sent them to dog-food canneries at four cents a pound. The photographs, taken in 1951 by Carson City photographer Gus Bundy, showed mustangers chasing horses in flatbed trucks across the alkali flats of Smoke Creek Desert north of Pyramid Lake Indian Reservation. From the bed of a truck running alongside the fleeing horses, cowboys roped them and threw out tires tied to the lassos. A horse dragged the tire for a mile or two until it became exhausted; by then, it was sometimes snorting blood. Captured, the horse's legs were chained, which made it easy to winch it into a stock truck for the trip to the cannery.

The 1959 federal law, which became known as the Wild Horse Annie law, was significantly different than its 1955 Nevada counterpart; the new law forbade the use of airplanes and motorized vehicles for rounding up horses on public lands—if the horses were unbranded. Ranchers were disconsolate, since all but a small fraction of

the land on which they fed livestock belonged to the public. They wanted to keep wild horse laws in the hands of the states, which they felt they could control. For its part, the BLM objected to the application of the law to public lands on the grounds that, without airplanes, the agency could not possibly control the horses.

Mustanging with airplanes continued in spite of federal law. Not only is the basin and range country of Nevada difficult to police, but ranchers and mustangers had a bagful of ruses at their disposal. If caught with wild horses and a smoking plane, they could claim that they were rounding up their own branded stock and "accidentally" captured some unbranded horses in the process. Or they could argue that they were merely gathering the offspring of animals they owned. Getting rid of unbranded mustangs was a minor problem, since there are ways to brand a horse and make a week-old burn appear as though it's been there for months. Though Nevada requires brand inspections before horses can be sold or moved any appreciable distance, this too was easily hurdled. Many inspectors were former cattlemen and therefore willing to go along with ranchers' wishes.

Wild Horse Annie was sorely aware of these difficulties, never more so than when her detective work pointed to widespread illegal mustanging. During one six-month period in the late 1960s, Nevada state records showed that more than two thousand unbranded horses were taken out of the state. Wild Horse Annie knew many of the violators on a first-name basis; she just couldn't summon enough airtight evidence to get any of them convicted.

In 1967 the BLM, aware of increasing public sympathy toward wild horses, included them in their multiple-use concept of public lands. The horses, along with cows, sheep, and wildlife, were to have more-or-less equal access to public grasses. Still, there was no easy way for the agency to implement its equal rights policy. The horses are not considered wildlife, either by state and federal wildlife agencies or by most environmental groups, and, even if they were, wildlife does not come under BLM jurisdiction. By the laws of most states, wild horses—so-called slicks or mavericks—belonged to states under estray laws. These laws gave the states ownership of the horses, even though they were born and lived their entire lives on land owned by the U.S. government. County commissioners continued to give ranchers the right to remove wild horses, the rationale being that the animals stole ranchers' grasses, water, salt, and, given enough time, would even rob them of their very livelihoods.

But Wild Horse Annie and her growing cadre of supporters didn't give up. In 1965, she and Helen Reilly, another Reno resident, formed the International Society for the Protection of Mustangs and Burros. Then, in 1971, Wild Horse Annie put together a sister organization called Wild Horse Organized Assistance (WHOA). The first society was primarily informational, the second surveillant. By the late 1960s, the celebrated horse cause had become truly national. People of all persuasions wanted to do battle for mustangs. Cattlemen became increasingly angry and defensive. They claimed that the wild horses belonged to them, and that the mustangs were not all that much different from cows: a renewable resource to be harvested for meat.

Shout as they might, the ranchers were no match for the pencil-and-crayon war waged by grammar school teachers and their tens of thousands of pre-teen students. Nor were ranchers a match for starched housewives and pearly-robed ladies beholden to childhood images of Bambi and Black Beauty. United States congressmen and senators suddenly got more pro-wild horse letters than they received on all other concerns combined. One senator alone allegedly received fourteen thousand letters, nine thousand of them from children. National media joined the cause. *Time,* for example, embellished history by describing the horses as "descendants for the most part of proud Andalusian horses brought to the New World by Spanish conquistadors 400 years ago . . . the only remnants that as recently as 1900 numbered in the millions." When ranchers said that mustangs were destroying the range, supporters of the horses countered with the same arguments ranchers themselves sometimes used. In the winter, horses break ice and keep open water holes, and they paw for feed and thus prevent cattle from starving. Their manure reseeds and fertilizes the range. Lawmakers couldn't ignore a few basic facts. The public lands, they learned, were home to hundreds of thousands of cattle and sheep who grazed for pocket change—ten to twenty dollars a year per sheep or cow. To boot, livestock in the critical wild horse areas of the West was only supplying one percent of the nation's beef needs and six percent of its lamb demands. One clear message imparted to legislators was appealing: remove all of the cattle and sheep from the nation's public lands and give them over to wildlife and wild horses and the nation's flesh consumers would hardly notice.

The wild horse cause even received support from what might, at first blush, seem a most unlikely person for such a crusade. For a couple of decades, Storey County, a barren little triangle of land with only one town of note—Virginia City—and a population of under a thousand people, has been home to the nation's most famous legal brothel, the Mustang Ranch. So important has the brothel been to the county that for many years the license to operate the house accounted for ten percent of the county's annual income. Long owned and run by the notorious Joe Conforte, the Mustang's forty or fifty prostitutes oblige the carnal desires of young boys and married men. It so happened that long before Joe Conforte knew of Wild Horse Annie, he had come up with his own concept of a mustang. His brothel logo is a barebreasted nymph whose lower half is a well-proportioned feral horse. Conforte was making so much money off his "working girls" and so much from the sale of decanters with his mustang cathouse logo on it that he offered to give all the proceeds from his porcelain sales to the wild horse cause. Wild Horse Annie, however wild in the eyes of ranchers, wanted no part of money spent by womanizing *Homo sapiens.*

Today, the Mustang Ranch and mustangs are easily brought together in the mind of an alert traveler on Interstate 80, even if one doesn't know about Wild Horse Annie. Less than ten minutes east of Reno is the unincorporated town of Mustang, and the turnoff to the Mustang Ranch. The town of Mustang—related in name only to the brothel—consists of a haphazard collection of thirty or so trailers, and the Mustang Lounge. The town and the lounge took their names from the wild horses that for decades grazed on the slopes of the surrounding mountains. In fact, the horses started claiming the water in the Truckee River and the grasses around the town of Mustang long before Conforte established his infamous business, and long before Wild Horse Annie happened upon her cause.

In January of 1970, Senator Clifford Hansen of Wyoming introduced a bill in Congress to authorize the Secretary of the Interior "to protect, manage, and control free-roaming horses and burros on public lands." By the early part of 1971, the bill had picked up strong support from the likes of Mike Mansfield, Frank Church, Mark Hatfield, and Henry Jackson. Support was so strong that more than fifty wild horse protection bills were introduced in Congress. Collectively, they were buttressed by a greater number of letters than was written on behalf of any other issue of the times, save U.S. involvement in Vietnam. On June 29, 1971, the Hansen bill passed the Senate without a dissenting vote. It galloped through the House without opposition, and on December 15, 1971, President Richard Nixon signed the bill into law. Passage of the Wild Free-Roaming Horse and Burro Act, as it was officially known, was Wild Horse Annie's brightest moment. Congress declared that: "Wild and free-roaming horses and burros are living symbols of the historic and pioneer spirit of the West; that they contribute to the diversity of life forms within the Nation and enrich the lives of the American people; and that these horses and burros are fast disappearing from the American scene. It is the policy of Congress that the wild free-roaming horses and burros shall be protected from capture, branding, harassment, or death; and to accomplish this they are to be considered in the area where presently found, as an integral part of the natural system of the public lands."

Congress made it the job of the Bureau of Land Management to act as custodian for all of the nation's wild horses and burros. The bureau could destroy old, sick, or lame animals, but it had no authority to dispose of them for commercial gain. It looked as though Wild Horse Annie had won, and perhaps she had—though she had only a couple of years to enjoy the fruits of her triumph on behalf of the horses. She died of cancer in 1977.

"Start looking for the trap and some vehicles," Ron said. His words were still lingering in the eddying air when I looked straight ahead into the middle distance and saw a stock truck, two toy-sized humans, and the long wing of the trap. With the gentleness of a father putting his first born in a crib, Ron landed in saltbrush and sage a hundred yards from the portable corrals. We unloaded the extra gasoline cans and began preparing for the chase.

The horse trap was on the north side of an unassuming hill dotted with matronly junipers. Just to the east was a draw and, beyond that, another rocky rise of more imposing dimensions. The theory behind the trap's location was simple. Several bands of wild horses, numbering perhaps eighty animals altogether, were known to feed and water not more than a mile to the south in a pastoral valley. Ron planned to move in on a small bunch, cut them away from the others, and then push them toward the draw. Once over the draw on the trap side, Bill Stewart, ex-BLMer and experienced roper, would intercept the horses from the east and drive them toward the trap. Galloping at full speed, the mustangs would see the trap wing and, with Stewart to the east of them, turn to the west and into the corral.

As I walked toward the wing of the trap, Bonnie Cox, wife of Woody Cox, the rancher claiming the horses, was on her knees smoothing dirt over the rope to the corral gate that she would pull once the horses were inside. Not far away, Bill Stewart was busy tying juniper branches onto the six-foot-high steel panels that made up the trap's sixty-yard long wing. Woody Cox, stumpy and rough-hewn, an imperishable character out of the Pecos frontier country, was doing the same. Stewart greeted me—we had met and talked before—and then began anxiously mumbling that he and Woody still didn't have the panels disguised well enough. "There's reflection off the bars and the horses can see it," he said. "They get spooked and turn away. A couple days ago we even rubbed mud on the panels to cut the glare. It didn't work for a darn. They turned on us and we lost them." The barrier of horizontal bars, an inch in diameter and about a foot apart, looked naked, all right. The panels had been painted with an off-green, high gloss kitchen paint; they were not intended for catching wary mustangs. So far as I could tell, no one had done much this day to cut down on the glare from the panels, perhaps because there was no mud at hand. Stewart tied on another piece or two of juniper with wire and shook his head. Irritation was writ large on his scarlet face. Wearing a white cowboy hat and brown-and-white fringed chaps, he walked over to Joe, his dark bay horse, and affectionately slapped him on the neck. He was anxious to run horses; to hell with fooling further with the trap.

We found mustangs in abundance in the adjacent valley. As we began to slope toward them, several husky necks rose to greet us and then the horses began to turn and run away from us. "They've been run too much before," Ron said. He quickly picked out a group of eight closest to the draw and zoomed in low to one side of them. Several turned in apparent confusion. Four came together, three joined in a separate powwow, one continued chomping away as if it hadn't eaten for several days. Ron moved closer and the four split up, circled, then snorted and joined the smaller gathering. Slowly, cautiously, in-decisively, all of them began moving away from the bruising, un-welcome engine clatter and wind rushes. Now Ron moved still closer and lower—a mere dozen feet above their sturdy bodies—then abruptly backed away. A handsome bay yearling decided to head for Virginia City, away from the others and away from us. Ron applied power and we pitched and dropped to remind the resister that his wayward behavior wasn't appreciated. The yearling got the message and scampered toward kin.

Ron pulled back to think over strategy. He put his yellow-eyed, red-tailed, quarter-of-a-million-dollar machine on an imaginary two-story landing pad where we sat for about thirty seconds. The horses had begun steadily marching in the general direction of the trap. They were not all of one mind, however; two of the bigger ones stopped to steal a couple of extra bites, then they turned back into the cluster-ing pack. The horses headed into a slight depression that led to higher ground and the draw. When the first of them neared the gap that would give them a full view of the trap wing, Ron turned on a blaring siren and came in low enough for both of us to take a good horse kick to the head. Those at the front of the galloping horde began sprinting over the lip of the draw. They headed right for the trap—or so it seemed for a few seconds. They needed an extra push, but there was no head wind, and this meant trouble. Without a head wind, Ron couldn't get the over-boosting maneuverability he needed to really crowd the horses.

Five of the mustangs were running right for Stewart, who had positioned himself on Joe behind a juniper about halfway between the draw and the trap wing. From above we could see him leaning forward in the saddle, anxiously waiting for the horses. In the lead were two bays, their manes aflame. Not far behind was a dark sorrel, another bay, and a red roan. As the leaders reached Stewart, he bolted out to meet them. He aimed to keep them between Joe and high ground, on a direct route toward the trap. Before the first horses had reached Stewart, however, he looked toward the draw and saw that those bringing up the rear were in disarray, starting to break to the east and away from the wing. Without a head wind as we came through the draw, we couldn't help. Stewart, knowing that he was now on his own, let out a hoot and a wailing holler, and he let loose with a threatening show of his lariat. Joe kicked high and swung to the left, then abruptly to the right; he seemed confused. Joe was confronted not with a recalcitrant cow but with a frightened phalanx of sweating horseflesh. The mustangs had smelled poisonous wind; instinctively they knew what to do. They darted for unobstructed high ground and blessed blue daylight. Stewart saw the futility of further pursuit and pulled up hard on Joe. Ron wouldn't give up as easily. We motored after the fleeing manes, but by this time our effort was as useless as an extra car horn in Manhattan rush hour.

Meanwhile, one cause evidently lost, Stewart turned and charged ahead to prevent more losses. Only two of the mustangs had gone into the trap without a hitch. Two more in the first wave had somehow gotten around Stewart and were now topping Freedom Hill. The fifth one paused and circled in befuddlement. Stewart gave chase, his hat went flying, and the bay—a real beauty with striking white leg warmers—made a beeline for the poorly disguised wing. Thinking it was seeing full daylight, the frightened critter hit the steel wing at full throttle, thudded to the ground, vaulted to its knees, and then, behind Stewart's yelping and flying circle of rope, ran into the trap. Three in and no more in view, Bonnie Cox pulled hard on the buried rope and the gate slammed shut. Bonnie stood up, wiped off her dusty glasses, then shook her head in apparent disgust. Three in the corral was nothing to brag about. Nothing at all.

Bill Stewart and I sat on a couple of boulders waiting for Ron's approach on the south side of the draw. Stewart was irritable, and it was time to take the pressure off his duodenum. "I got to have things going my way. That panel wing was placed at too great an angle. It gave the horses a chance to turn around and go back. It shouldn't have been done that way. It should have been moved. Right now it's down to where guys like Woody need some help on these roundups. Not that I know it all, but I learn something every day. Just like the other day I was arguing with Woody about that pen back there. I said to him, okay, it don't make any difference to me, you go ahead. But I've never rigged 'em that way. This wing, Woody wanted to bring it straight out. Right out into the open!" Stewart picked up a twig and scratched a picture in the dirt. "You do it like that and those horses come in and see that wing and they'll just run around and head right back out. And Woody, he said, 'Oh no, oh no! It'll work.' And I said, 'Buddy, I gave into you this morning, I'm not giving into you now. I got several thousand dollars wrapped up in these panels and they're here on borrowed money. I gotta get it out of these horses and the bank says I only got ninety days to do it. Those wings are going where I want 'em.' You know, I just had to be hard with him and, by God, he took it."

The acid in his stomach neutralized and still no indication that Ron was nearby with horses, Stewart decided to talk about his favorite subject. "I was born and raised a cowboy and I don't wanna be nothin' else," he began. "I caught my first horses in 1936 up there in the Bluewing Mountains, ninety miles northeast of Reno. Off and on, I been running 'em ever since. I used to go out and catch a couple of horses if I wanted to go to a rodeo and didn't have enough money."

He went on to say that when he was young he rode bareback, and that nothing was a greater thrill than sitting atop a fuming bull. From there he moved on to professional calf roping and bulldogging. When his bones and muscles could no longer take the beating, he turned to team roping and, in 1969, was on the Nevada State Roping Championship team. Now, beyond the age of professional competition, he recalled his past accomplishments. "I've roped as high as six mustangs in one day," he said proudly. "Back then I'd put a sideline on them, go home to camp, and then come back the next day and pick 'em up."

Bill Stewart never owned a cattle outfit himself, but he's worked for several. For a number of years he was a foreman for the Horseshoe Cattle Company, an operation that ran six thousand cows. "We'd run about one hundred to one hundred twenty-five head of horses in the cavvy, that's how many cows we had to take care of. At one time I did a lot of flying for the ranch. I got a commercial pilot's license, and I'd take the door off and put a guy in the back seat and fill the plane up with salt blocks. I could salt that whole ranch in about two hours."

In the 1950s, Bill Stewart and his family moved into Reno, because it was only forty miles from the ranch. He hated every minute of city living. "That town really bothered me," he says. "I hate to admit it, but it embarrassed me to pack the groceries from the car to the house. It really embarrassed me. My wife wanted me to mow the lawn and I said, 'Honey, I ain't gonna mow that damn lawn. Just look at all those people looking out through their windows. Whatdayou think they'll think of me?' And my wife said, 'Boy, you been in the bush too long.' And I said, 'Yeah, and that's just where I belong.'"

Stewart finally quit the Horseshoe Cattle Company and moved to Elko where he bought a hundred head of cattle and began renting roping cows to rodeos. He liked working the rodeo circuit, but by 1961 he had tired of the uncertain paychecks. When he heard that the BLM was looking for a range rider in its Carson City office, he applied and got the job. Then in 1971, the BLM was suddenly told by Congress that it had the job of managing the nation's wild horses. Stewart showed interest and was immediately appointed one of the bureau's first "wild horse specialists."

Before the end of his first year in the wild horse program, he was enlisted to work in a much-heralded emergency roundup in Montana's Pryor Mountains. That winter had been a particularly bad one in Montana and Wyoming, and without the roundup dozens of wild horses would have starved to death. Stewart and a few other BLMers spent eighteen bitter-cold days on horseback gathering forty-five mustangs. After the horses were nursed back to health, they were given to Montana's Crow Indians. For his effort, Stewart received a Department of Interior citation. Reflecting on the experience, he says, "We roped 'em all and tattooed 'em all and from then on Wild Horse Annie and I were good friends." A long moment passes before he adds, "Wild Horse Annie had a lot of good points and I had a lot of respect for her and what she was doing. But I gotta tell you, there are cattlemen in this country who are glad she's dead."

In 1975, Bill Stewart was put in charge of Palomino Valley, where all of Nevada's captured wild horses and burros are branded and put up for adoption. The job wasn't particularly to his liking, and two years later when the state decided it needed its own wild horse round-up crew, he jumped at the chance to get back on a horse and use his riata. From August of 1978 until his retirement in May of 1982, he called the shots on when and where mustangs were to be captured. Whether it was chasing the horses in a helicopter or pushing them the last hundred yards into a carefully concealed trap, or cajoling them into a stock truck, he loved every minute of it. Even now, at that age when many men think of spending their days on a sun deck in the Caribbean, Bill Stewart dreams of chasing mustangs. As he likes to say, "It's got to be a lot hotter than the hinges of hell to keep me away from 'em."

"We got a dead one," Bonnie said as Stewart and I neared the trap's corral. It was the same white-stocking mare that had hit the trap wing earlier. Bonnie said that as soon as the mare got inside the corral she continued charging at the panels. She did it three, four, five times . . . Bonnie stopped counting.

There was no doubt about how the mare had died. One horizontal bar, about three feet off the ground, was splattered with blood. The mare lay two feet away, her eyes wide open, glistening prismatically in the sharp light. Her neck was bent like an arthritic ladle; near the jugular groove was a lump the size and shape of a swagger bag. Bonnie crawled inside the corral and went over and knelt beside the mare. She ran her fingers through the dusty mane. She carefully straighted out the forelock. Bonnie seemed to be reassuring herself that there was no possible sign of life, nothing that could be done. Bonnie's face was long and private.

The two other mustangs were huddled in a brushy corner on the other side of the corral, next to the gate. They were small, lumpy, dull, unwelcome counterpoints to Everyperson's romantic image of spirit-ed, enchanted mustangs. I stared at them. Then, without apparent provocation, their hips touched and a sharp whinny pierced the pure air. Their rumps and rear legs thundered skyward, as if they had been stabbed with a charged prod. Slowly their heads came together and they did a quick two-step. They wouldn't turn toward the corpse.

Bill Stewart and I stood outside the corral. Neither of us said a thing. Finally, Stewart moved around to the near side of the warm, rotund body, glanced at it, then said, "That's the breaks of the game. Don't like it, but nothing I can do about it." He turned away and walked east, toward the rise where earlier the horses had escaped. He walked a dozen yards before stopping dead in his tracks. Wheeling around, he looked in the direction of the dead mare; he spoke as if talking to her. "Every time. Every time it's got this instinct to duck its head when it sees it's gonna hit something. Don't like it, but nothing I can do about it."

I couldn't shake Stewart's words, and more than a year later on a visit to a BLM adoption center near Susanville, California, I recalled the resigned hurt in his face. The same day, one hundred thirty-eight good-sized California mustangs were trucked into Susanville and put out on ten acres of pasture. Four of them broke their necks trying to charge through fences.

Woody began chasing the two captured mustangs into the outer corral. They kicked and resisted and banged violently into the panels before scampering into the outer catch pen. The gate secured, Woody removed two panels near the fallen mare and tied a rope around her hind legs below the hocks. He backed up his stock truck, shoved it into gear, and the mare's dead weight kicked up dust and broke clumps of rabbitbrush as she was pulled outside the corral. Then Woody got on a small bulldozer that had been used to plow a road so the trap panels could be brought in. He tied the loose end of the rope on the back hoe and crunching forward, the heavy load in tow, headed for a hillock west of the parked helicopter.

"She was the prettiest, the best we got," Bonnie said mournfully. "I bet she was four years old. She was the nicest one and would have made a real nice horse for somebody." Bonnie had a rich, deep tan, a shaggy ponytail, the nails of a grubby archeologist.

While all of us waited in silence for Woody to return, I was reminded of a conversation I'd had the previous year in Elko with a helicopter pilot by the name of Ted McBride. By Ron Crowe's standards, McBride's mustanging experience is modest. At most, he's coralled a thousand mustangs over the years. McBride believes that the most successful airborne chasers are those who use well-camouflaged and craftily hidden traps. "A mediocre pilot can put horses in a damn good trap, but a damn good pilot can't put squat in a bad trap," he says. Another McBride opinion is that using a helicopter to chase horses "is one hell of a lot safer than that Supercub I used in the '50s. It was under-powered, I was inexperienced, and I was just another one of those crazy Sunday pilots out trying to kill myself. I was so busy running horses I didn't fly the airplane. There were others like me, too. About the only thing we could do in a fixed-wing job was wear them down and get them close to the trap where the cowboys could take over. The good pilots in this country you can count on one hand, and I'll cut off three fingers before you start counting." When talking about wild horses, McBride likes to emphasize one of Murphy's laws: If anything can break when you're chasing horses, it's certain to break. "One hundred percent of the time," he says with the smile of experience. "Running horses is hard on helicopters, hard on good horses, hard on pickups, and real hard on fences. You've probably heard the joke that ranchers tell," he says laughing. 'Well,' they say, 'it's about time to run horses because it's time to repair the fences.' Wild horses don't respect wire and they don't respect posts. I've heard of wild horses going through five or six fences in a single chase. Why, I've seen them go through fences just ass over teakettle, and then jump up and go right through the next one. I've seen them go through eight fences in half a mile. If you ran a saddle horse through the same fences you'd have to shoot him. Don't ask me how they do it, but they do it."

Ten minutes later, Woody returned. He had dumped the carcass in a ravine. "For coyotes," he said evenly. "There's lots of them around. They need them horses. They'll do good on her."

Woody suggested that all of us sit under some trees and have something to drink. He pulled a beer out of a cooler and held it high. No one besides Woody seemed interested. Ron and Woody, but not Stewart, said they thought the wind would die down or reverse direction in a half an hour or so and we could resume chasing horses.

Bill Stewart began counting how many mustangs had been caught since the roundup first began a week earlier. "That first batch weighed six hundred thirty-one pounds on average. They were small, sure to go to the pet factory. But really, that's a pretty good weight. Some say horses were weighing five hundred pounds in here thirty years ago, but I don't know about that." The Virginia Hills horses, Stewart told us, are some of the smallest in the state. Elsewhere in Nevada mustangs weigh eight hundred or nine hundred pounds. The really big wild horses—eleven hundred to twelve hundred pounds— are found in Wyoming's Red Desert, where ranchers have culled small stallions and put out Belgians and Percherons to take over the harems. Generally, feral horses are smaller than domestic ones. Their size-genes have not been constantly manipulated by humans. Also, their diets are often deficient in vitamins and minerals, salt and water are not always available in sufficient quantity, and at times they just plain don't get enough to eat.

Stewart had agreed to help the Coxes with the roundup for a percentage of the sales price. He agreed because he had a pretty good idea how much he could make. From his BLM days, still as fresh as a fragrant rose in his mind, he knew how many horses were in the area. In 1978, while still with the BLM, Stewart had counted the Virginia Hills horses and came up with a figure close to seven hundred. "In these parts, where Woody runs 'em, I counted four hundred four years ago," he said. "Just a few days ago I was up in a plane and saw three hundred seventy-five. But I know we missed a bunch, twenty or twenty-five percent anyway."

Woody Cox had been running horses on his 102 Ranch for four years when he asked Stewart to help him. "We lease our land from Curtis-Wright," Woody said. He motioned to the distant mountains, toward Virginia City, then in the opposite direction. He explained that when Curtis-Wright bought the Virginia Hills' land their property line extended from highways 80 to 50 and from 395 to 95A. "That's how come we run them, 'cause they're on private ground."

In 1957, the Curtis-Wright Corporation bought seventy-seven thousand acres north of Virginia City. The land was to be used as a testing ground for rocket engines. When the plans went awry—they still have not done any testing in the Virginia Hills—they leased parcels to ranchers for grazing cattle. They also gave the ranchers permission to graze horses free of charge. The Curtis-Wright rationale was that by eating the grasses, the horses would reduce fire hazard.

By the mid-1970s, a couple of ranchers leasing Curtis-Wright land began to feel cheated; someone else's wild horses were taking grasses that they needed for their cows. They complained to Curtis-Wright that there had to be some removal. But by a curious twist of law and logic, the BLM entered the picture and forbade anyone in the Virginia Hills to round up horses. The BLM's jurisdiction is limited to public lands, but somehow the agency had gotten itself into the position of speaking for rights on private lands. It took until late 1976 for the bureau to leave Curtis-Wright to its own business. The way cleared, Curtis-Wright went along with the complaining ranchers and agreed that a major harvest of horses was necessary. Local mustangers promptly put in ownership claims. One of the ranchers stepped forward to proclaim that he had fifteen hundred "domestic" horses in the Virginia Hills. He said that he'd had them out there for more than twenty-five years.

Billy Boegle is another rancher who leases land from Curtis-Wright. When I talked with him at his ranch north of Carson City, he was amiable and eager to convince me that his methods of rounding up horses are perfectly humane. He says that he likes to gather them every couple of years to cull out the older ones and keep the population at a reasonable level. Dead Horse Basin, an above-ground cemetery of skulls and bones that acquired its name from past tragedies, is a place he likes to take anyone who doubts the need for culling. Boegle says that whenever he does a roundup he takes a Nevada State Brand Inspector along to certify that everything is above board. He has sold over five hundred Virginia Hills wild horses since 1976. Any buyer is okay with him, though he prefers to sell his catch in California, where there are no brand inspection laws and where there is less likelihood of bad publicity.

Boegle says that he has been putting his own good studs into the Virginia Hills since the 1950s. In 1969, he started releasing some large work stock, "to put more meat on the hoof." Pounds and market prices are central concerns to Billy Boegle. "For these Virginia Hills horses you can get ten to twenty-four cents in the summer, near to fifty cents in the winter when there's a big demand overseas in Belgium and France. They go out of here through Nebraska. Sometimes they send them down to the Texas rendering plant. When they weigh over eight hundred pounds they go for human consumption and you get better prices. We don't get much of that bigger stuff around here. I'm ahead if I can sell my small stuff to private parties."

I didn't get a chance to ask Boegle what role he had played in the 1978 and 1979 roundups that had netted several hundred horses, most of which were sent to a slaughterhouse in Olympia, Washington. But he might not have wanted to talk about the roundups, which would have led into his ongoing battle with Gerry Olsen.

Gerry Olsen had long dreamed of living in a high-beamed, knotty pine country house away from the maddening crowd. When she and her husband discovered the Virginia Hills with its abundant pinyon pine, high sage, winterfat, and unpaved roads that meander among orange and sepia hills, they figured they had at last found the perfect retirement haven. They bought an acre of land, put in a septic tank, dug a well, and began work on their home. The Olsens bought their property from the Virginia City Highlands Association, who had bought the land from Curtis-Wright when it was finally decided that the Virginia Hills would probably never be used for rocket testing. Five subdivisions were created, with lots ranging from one to forty acres.

The Olsens moved into their dream home shortly after the Virginia Hills roundups of 1979, and Gerry soon heard a variety of unflattering tales about the gatherings. Even though she admits to only having seen one, and that was a "very small one conducted by the BLM that was very, very nicely done," she's not convinced that most roundups are humane. She wants to place a lot of blame for inhumane treatment on the helicopter. She and her subdivision friends would like to see the roundups take place on horseback. I asked if she understood that chasing mustangs on horseback was dangerous to riders and frequently crippled good saddle horses, to say nothing of being downright difficult and inefficient. Her response to me was, "Well, then, it would be better to use water traps." Perhaps—if enough of the water that horses use can be guarded.

What really bothers Gerry about roundups in the Virginia Hills, or anywhere for that matter, is the prospect that the horses will be slaughtered for food. She is unabashedly sentimental about the animals in this regard. "I see them out here and I really get attached to them and I don't want them going to a rendering plant. That's the cruelest thing of all."

In 1979, Gerry began organizing the Virginia Range Wildlife Protection Association. Its original purpose was to protect the mustangs, "because I didn't think the ranchers had a right to do those roundups, even if they were doing it on their own property. Some of those horses belonged to us." The association was backed by Virginia Hills developers, who recognized that the horses would help sell parcels of land. By 1981, the association had a voluntary deputized range rider. Once a week he rode in the hills trying to protect a few eagles and a lone mountain lion, advising those with four-wheel drives to take their destructive machines elsewhere, and telling people that the wild horses were legitimate wildlife just like deer and antelope. As enthusiasm for the association grew—by 1982 nine hundred people had signed a petition stating that the Virginia Hills horses ought to be left alone—new reasons for protecting the horses were put forth. The association claimed that the horses will reduce the likelihood of a devastating fire, and that "herds of them running wild will be a big selling point to tourists."

Virginia Hills is largely unfenced, and cows and wild horses move freely between public and private holdings. Aware of how easy it is for mustangers to push horses from the subdivision lands onto Curtis-Wright property or their own ranches, Gerry and friends began the fatiguing task of posting signs every seven hundred feet around the entire perimeter of the housing development. The signs read: "Warning—Private Property. No hunting. No trapping. No removal of horses or any other wildlife without permission of owner." Not content with this easily frustrated maneuver, Gerry did some historical research. She discovered an old estray law enacted by the Nevada State Legislature that allows people to gather or shoot wild horses. Gerry hired a lawyer to get a ruling on whether or not the law applies equally to public and private land. It does, she was informed, but the mustanger must post a bond and get permission from the county commissioner before going after the horses. She thought she might be able to catch the mustangers on a technicality. But Boegle was aware of the estray law. As early as 1975 he had brought the regulations to the attention of the Storey County sheriff. For the moment, Gerry has had to settle for a resolution from the Storey County Board of Commissioners which recognizes the concerns of the Virginia Range Wildlife Protection Association—"its dedication to the preservation of the wild horse population"—and notes that "no gathering of wild horses will be permitted without the approval of the Board of County Commissioners." Roundups must be humane, and Gerry and anyone else is guaranteed the right to observe them.

After Gerry and I talked about horses in her sunny ridgetop home, she was eager to show me how far she'd come from her citified ways. She brought out a book of desert plants and pointed to those she could identify in the surrounding hills, then took me into her garden for a tour of well-nurtured favorites: wild currants, squaw tea, wild rye, and that favorite among wild horses—Indian ricegrass. With an infectious smile, she explained that one of the decisive reasons for moving to the Virginia Hills was the wild horses. "Here's this animal that's always been man's companion. Here he's also wild and so beautiful to see. There's something special about a horse without a bridle on it, wouldn't you agree?"

It was easy to do so.

She pointed to the valley below, and said, "Once I was going over there and they came right up to me. They were interested, they were curious, and they were facing me! It was so exciting seeing a wild animal that close!" Gerry's eyes sparkled and her cheeks suddenly looked incandescent.

Gerry believes that the Virginia Hills horses, like the rest of Nevada's wild horses, can be traced back to original Spanish mustangs. She is unmoved by ranchers or others who take issue with her on this, equally unmoved by reminders that Wild Horse Annie felt that most, if not all, of the state's horses resulted from the doings of nineteenth-century ranchers and miners. Gerry smiles sweetly when told that Tony Amaral, a long-time Carson City resident and respected student of Nevada's history and horses, once wrote that, "The wild horses that live in the hills less than fifteen miles from Nevada's state capital, and east and north and south in Nevada's mountains and deserts, are no more representative of the Spanish horses introduced into America in the fifteenth century than current cowboys are representative, in soul and substance, of the old-time range riders."

I had the feeling that Gerry wouldn't have been moved were I to have mentioned that much of the popular misconception about the origin of wild horses derives from the word "mustang." Mustang is a corrup-

tion of the thirteenth-century Castilian word, *mesteño*, which referred to stray sheep belonging to a stockman's organization called
Mesta. Later, in the American Southwest of the eighteenth century,
mesteño became a handy word for a wild horse, irrespective of how
much Spanish blood it had.

Nevada ranchers, and a great many people who now have nothing
to do with ranching, insist that it was their forefathers who let loose
horses that became Nevada's mustangs. And, on the whole—obvious
self-interests and prejudices of the ranchers notwithstanding—they
may well be right. The journals of early explorers who traveled
through Nevada rarely mention sighting horses. Jedediah Smith, a
trapper and mountain man who journeyed through Nevada in 1826
and 1827 and is considered to be among the first white explorers of
the state, skirted the southern edge of Walker Lake in June of 1827.
There he saw "considerable horse sign." But other than this, he
apparently saw precious few signs of horses in Nevada. Nor did he see
much wildlife. After a twenty-two day trip that took him to the
southwest corner of the Great Salt Lake, he noted, "We had but one
horse and mule remaining, which were so feeble and poor that they
could scarce carry the little camp equipage which I had along. The
balance of my horses I was compelled to eat."

Milton Sublette, a trapper who worked the Humboldt River area in
1831, reported finding no wild game to speak of, and his party was
forced to eat beavers they had trapped, which made them sick. Later,
on a northward trek to the Snake River, Sublette and his men could
find so little food that they had to subsist on ants, crickets, and
puddings made from the blood of their pack animals. Another Nevada
explorer of the 1830s, Zenas Leonard, wrote in his journal that, "We
found the country very poor and almost without game except for some
bighorn sheep, some antelope and rabbits." With the exception of
rabbits and ducks, the Leonard party took no fresh meat. Captain
John C. Fremont, who entered Nevada from Oregon in 1843 and
journeyed through the western and eastern parts of the state, mentioned seeing waterfowl, sage hen, rabbits, bighorn sheep, and antelope. And some horse tracks—but no horses. Like others, including
many early trappers who spent a lot of time in the state and tried to
live off the land, Fremont's reports give every indication that game
was scarce, and wild horses even scarcer. If wild horses were there, it
is almost certain that explorers would have killed them for food rather
than shoot their saddle horses to avoid starvation.

None of this is to deny that there may well have been isolated
populations of mustangs in Nevada in the early nineteenth century,
and a few of them might well have had a pinch of Spanish blood in

them. But these few were nothing compared to the large numbers of horses brought into the state by cattlemen and sheepmen from the 1860s onward. Even before cattle and sheep herds grew to record numbers, there were large horse ranches. For example, by the 1870s the Button brothers in Humboldt County were reputed to be running as many as six thousand head of horses on the open range.

From the 1860s until the early 1890s, Nevada cattlemen raised little hay for winter; they turned their extra horses out to fend for themselves. Then, twice a year, in the spring and fall, cows and horses alike were rounded up. Or rather, when it came to horses, they rounded up what they could. In some parts of Nevada, horse gatherings took place infrequently. About 1870, domestic horses were turned loose in the Sheep Creek Mountains north of Battle Mountain by J. A. Blossom, a rancher and miner. Virtually no attempt was made to capture them until the last years of the nineteenth century, by which time their numbers had grown to more than two thousand. By this time, Blossom's son decided to profit from the mustangs. He sold many of them to the German army.

In the Diamond Valley in eastern Nevada, ranchers were raising horses as early as the 1880s. The better ones were broken and used on the ranch, a few were sold to rodeos, and good-looking mares and branded colts were returned to the unfenced range. Then, as in successive generations, mares only had three uses: for racing, for breeding, and as a remuda mare to keep saddle horses together. Now and then some of the horses were sent by rail to the West Coast, there to be used for pulling stages and freight wagons. Hundreds were shipped to San Francisco to help clean up after the 1906 earthquake.

South of the Diamond Valley in Nye County, ranchers and farmers joined together to purchase draft studs—Belgians, Clydesdales, Percherons, Shires—which were then crossed with ranch horses, and mustangs. When the crosses produced smaller, lighter-framed animals, they became saddle horses, which, when not in use, were let loose. In the 1920s and '30s, the offspring of desirable army remount stallions joined free-roaming horses. The remount program put thoroughbred genes into the mustang populations. As the government program began to demonstrate that quality light horses were valuable, more and more ranchers began to control the genetic composition of herds by shooting mustang studs and putting out better-quality stallions.

After the devastating winter of 1889–90, when more than a quarter of a million livestock in the state was lost (out of some seven hundred thousand cattle and four hundred thousand sheep), larger outfits broke up and smaller ones began to appear. Ranchers started growing hay that would carry them through the winter. The haying was done with horses, which were kept in a pasture. When they weren't being used, they ran free on the open range. A decent-sized ranch might have as many as seventy-five head of saddle and ranch horses. Whatever the number, some so-called ranch horses were never rounded up. Furthermore, because many buckaroos in those days didn't think that a horse was an adult until it was six or seven years old, a stallion had ample opportunity to lose domestic habits and reproduce. Mares, the factor limiting growth of herds, also had plenty of time to prove their biological worth because no self-respecting cowboy wanted to ride them.

When mines closed in great numbers in the 1880s and 1890s, hundreds of draft stock that couldn't be sold were put on the range and more or less forgotten. The Great Depression was another boom time for Nevada's wild horse populations. People tried to find buyers for their stock, but many couldn't, and just let them loose. After the 1934 Taylor Grazing Act, ranchers needed permits to graze horses on the public domain, but didn't make much of an effort to tell the Grazing Service what they thought they had. The agency did little to find out. And then around the time of the Second World War, when ranchers acquired trucks and tractors for haying, whole teams of draft horses were left on the unfenced range. For all of these unfettered critters, the only predator besides man was the occasional mountain lion.

Today, many Nevada cattlemen will offer their personalized version of this brief history. They will repeatedly emphasize that their forefathers are responsible for wild horses, that they have always managed them, and that they have always wanted them there. They wanted them on the range because, on the whole, they didn't have to pay for the feed they were eating, and because the horses were another payday. Sometimes a pretty darn good one. On a large outfit, the income from a mustang harvest every so often was enough to buy a new car or pay the kids' college bills. It's a damn shame, the ranchers now say, that the government has to interfere and take away their income and replace one system of management with a less efficient, more costly one.

Ron Crowe and Bill Stewart got up to check the wind direction. They quickly returned with the bad news. Woody shook his head, took a sip of beer, then said he wasn't satisfied that they'd chosen the best place to run horses. "We shoulda done more like we did the other day when we got twenty-eight." He paused and pointed west. "Well, you know there's a spring up at the end of that canyon. That's where we should be runnin'. Way up at the end of that canyon. There's a mine there where they used to make whiskey from that spring. They was makin' it right back in the old days, the Virginia City days, you know." He laughed.

"We started too late, that was the problem," Stewart asserted. "If we'da started earlier like I wanted we'da been okay."

"Maybe, I don't know," Woody replied.

There was a long pause. Then Stewart began describing how easy it is to catch mustangs if you're good with a rope. "The best time is after they have tanked up, but you gotta be careful. You can kill too many. Lots of 'em don't drink every day, they only come in to tank up every other day. If you really want to rope 'em after they've tanked up and they're fat all over from water, then you have to rope 'em and go with 'em."

Woody said, "We've roped horses at a lot of water holes and never lost more'n one or two."

Stewart frowned, like he didn't believe him.

"When we rope them we don't stop them cold," Woody added.

Stewart said, "Let me tell you about one night when I caught thirty-one head of horses at one trap, and I said 'god dang,' 'cause I'd never laid at a water trap all night like that before. I laid there at an old line shack and those horses was all lined up like a bunch of China-men. This was five o'clock in the morning and it was getting daylight out and they was just standing there. Finally this lead horse couldn't stand it any longer and he went in and started drinking and then another horse went in and bumped him and he drank. All it took was for that first horse to go in and that did it. All the rest of 'em went in. But before that lead one goes in, they'll sit out there and starve themselves of water if they think something's not right. They don't just come off the mountain and walk in a trap. They're coyote, they're real coyote.

"I don't like them traps with a trip wire," Stewart continued. "Three or four horses go in and the first one that trips it, the door slams shut and you don't get all of 'em. And the worst of it is you'll have a mare inside and her colt outside. So I don't like 'em. I like to use willow fingers for the gate. You take willows about the size of your finger and put three on each side and run 'em through each other and when the horses go through 'em, that'll hold 'em in there. Oh, they'll fight 'em. I went out to this water trap and we had fifteen head and they had all the willows bit in two. There wasn't one willow holding 'em in, no sir boy. When I came up over the hill and I saw 'em, I made a beeline for the trap, and I got fifteen. I love to water trap. Sometimes I'd rather water trap horses than fly them with a helicopter. It's a bigger challenge."

Stewart's enthusiasm for water trapping reminded me of a story I'd heard a month earlier from Ed Depaoli, a BLMer who catches wild horses in the high desert north of Lakeview, Oregon. "I'll tell you somethin' that's damn strange," DePaoli said. "Last year we got forty-two head in one clatter. Don't ask me why. All I know is we gotta let it happen when it's ready to happen." He explained that for days horses came to drink, but invariably three or four stayed outside while the others tanked up. Then when the stragglers did enter the trap, several from the bunch that had been drinking left. "You don't want to catch part of them and spook the others. You might get those spooked ones eventually, but they'd be the last to go in to drink and you could be there forever." From his bosses, Depaoli began catching hell: they wanted results, and they weren't mollified when he explained to them that "spookin' and lettin' a few get away" was a bad idea. "You got to be patient. Then this one evening, the last day we was gonna be out there, I couldn't believe it. They came just before sundown. That's when things start happening, when it starts cooling off. When it's hot, up around a hundred, the horses just stand there like they're dead. Anyway, I was out there that night and I knew something was going to happen because the horses were bunched up and a few would come and a few would go. But then I'll be goddamn, they just started spinnin' in and they just kept comin'. I think we had three or four head stay outside and not come in. We were saying to ourselves, to hell with it. We got enough in there anyway. Then the rest went in and we caught every one. I looked down and saw all those horses in that trap, all forty-two of them, and I said, 'Jesus Christ, we gotta be careful. We got lots of horses.' Did we ever have lots of horses!"

Bonnie, acting like she heard Stewart's tales every time they took a break during roundups, asked Ron Crowe what his all-time record catch was for a single day.

"Not much. Forty-five, I think." He said he had pushed horses into small traps and traps without wings, with and without the aid of cowboys, occasionally with "Judas" or tame "traitor" horses. A couple of times he had tried using a relay system, pushing horses ten to fifteen miles a day for two or three days to get them closer to a well-placed trap. But he found that the method didn't work very well, that when he returned the next day most of the horses had gone home. "They say these horses aren't territorial, but that doesn't match with what I've seen."

Stewart threw in the tidbit that in the early 1970s he had special permission from the U.S. Attorney to chase mustangs in his green-nosed Supercub. To get the horses running in the right direction, he backfired the plane or shot near them with a .410. "The worst thing was I'd go into the trees with ten of 'em and come out the other side with five. Half the time I couldn't get 'em out. I'd even drag rocks over the top of the trees to get 'em out to where I could push 'em.

"Some day I'm gonna build me a quarter-mile trap for these horses," Stewart said, starting to laugh. "It's gonna be built like a big inflatable balloon. We'll drive the horses right into the middle of it and then when the last one gets in the middle, I'm going to pull the plug, just like you do in these life rafts. That thing is going to go *Pheee-wheew!* right up around 'em. We got 'em! We got 'em! Then we'll put 'em up in the corral and deflate it and start all over again." He chortled and his face turned veiny, rosy pink.

We all laughed.

"That's like those Indians," Woody said. "They go in the mountains with a helicopter and a big net and drop it down on the ground. The horses come running down and the helicopter lifts them up in the air. Or something like that. Or instead, the helicopter brings them horses down low and the Indians start chasing them real hard."

"That's right," Bonnie chimed in. "They go hell bent for leather. You want an insurance policy before you go out with them."

Stewart said, "They like that plastic stuff with nylon in it, I don't know what it's called. They lay it down in sagebrush so it don't show. They bring those horses down hard and yell all the time. Once they get 'em close to the trapwings, they're gonna get 'em. A guy gets in a truck and guns it and that plastic stuff comes shooting out of the ground and they got 'em."

Just about everyone in Nevada with an interest in wild horses has stories to tell about the Pyramid Lake Paiute Indians and their sporting ways with mustangs. Since the federal law against chasing and catching wild horses doesn't apply to those on private land or on reservations, Indians are free to treat them as fancy dictates. Now and again during the warmer months, when the Indians are not on the rodeo circuit and have nothing better to do, someone will decide it's time to have a little fun. Perhaps as many as fifteen or twenty of the reservation locals will load up their trucks with saddle horses and cowboy gear and head north from their home in Nixon to the Smoke

Roping mustangs is a popular Paiute diversion.

Creek Desert and the nearby mountains in the northern reaches of the reservation. As often as not, the Indians run down the mustangs on horseback. The Indians will rope a few, they'll hobble them for a good look at their catch, and then they'll talk about the thrill of it all before releasing them. Occasionally, the Paiutes will rent a helicopter for a day, have the pilot chase the horses down from higher and rougher terrain and then, the wild ones thoroughly bushed, they'll chase and rope them.

By all accounts, the mustangs the Paiutes catch are small, similar in size to the Virginia Hills horses. Often they don't weigh much over five hundred or six hundred pounds. The Indians claim there is a lot of inbreeding among them and that as a result, "they're not much good except for kid ponies." A few of the very best are sometimes used for bucking horses in Indian rodeos. A few are sold, but not many. As the Indians tell it, by the time they pay for a brand inspection and the auction yard commission, it's hardly worth their trouble to catch and transport the horses. Still, there is some selling going on. Just before an Indian rodeo, Paiutes in need of quick cash will catch three or four mustangs and take them to the auction-yard in Fallon. From time to time, horse buyers from California come onto the reservation to buy anything available. Invariably, the horses go to rendering plants.

Chasing wild horses for sport is by no means solely an Indian pastime. A couple of months before meeting Ron Crowe and Bill Stewart, I talked with Sherm Tolbert, who owns Delta Livestock and Auction in Delta, Utah. With the biggest livestock auction business between Salt Lake City and Fallon, Tolbert and his sons handle about one hundred seventy-five horses a week. Most of them go to rendering plants. Although the Tolberts feel there are presently too many wild horses in Nevada, and the slaughterhouses are the right way to dispose of those that have poor conformation or cannot be adopted, Sherm Tolbert has a long-abiding interest in the animals.

Tolbert's recall takes him back to 1921 and his boyhood days around Delta when he accompanied two of his uncles on an extended roundup that netted several hundred horses. "There were big ones, small ones, and all manner of ugly ones," he says. "Why, there were so many back then you could stand around out here and move your arm in a big circle and you'd see more horses than you'll ever see in your life."

Sherm Tolbert may exaggerate now and again, but his memory is pretty good. Others who were around at that time estimate that more than ten thousand mustangs once roamed the Sevier, the Black

Rock, and the Escalante deserts, and the Confusion and Wah Wah mountains of western and southwestern Utah. Farther south, between Cedar City and the north rim of the Grand Canyon, there may have been another fifteen thousand wild horses.

Some mustangers trapped horses by building corrals around springs and water holes. They hid in brush or lay in a hole. After several horses entered the corral they'd quickly close the gate. Others used a relay system, several riders taking turns chasing a herd. Some rode bareback, with only a surcingle to hang on to. This was done to eliminate weight and make it easier for the mustanger to get free from a falling horse. It was not uncommon for their saddle horses to step in badger holes, breaking a leg and injuring the rider. Tolbert says he went after mustangs on his own, sometimes following a small band for several days. He brought them into Delta and sold them for a dollar or two each, anything he could get for them. He claims that despite their numbers, they were difficult to catch. Or, as he puts it, "That Zane Grey stuff where you'd just shoot the stallion and the mares would run into the corral, that was just bull fiction."

One of Tolbert's stronger opinions is that it is a lot harder to catch a wild mare than a stallion. A favorite story of his concerns a big grey mare—"nine hundred eighty pounds, not over one thousand pounds, they put out a lot of Percherons in this country"—that it took him almost two days to track down and rope. She was, he says with a fisherman's shifting brow, so wild that the best thing he could imagine doing with her was to give her to a close friend as a joke. When his friend saw what he had taken, he decided the only way to get a saddle on her was by lowering it down from an over-hanging apple tree. As Tolbert tells it, laughing a hearty line between each word, the first couple of times the saddle touched the mare's back, she kicked it high into the sky. The friend persisted with his apple tree hoist, and finally the mare took the saddle. Then he decided to get on the feisty mustang the same way the saddle had gone on. The mare roared and kicked up a storm and by Tolbert's estimation, it was as good a rodeo show as you're ever likely to see. Before it was all over the cowboy went flying, his left shoulder looked like it had been hit by a log splitter, and the flank billet had been broken beyond repair. Knowing that he had more on his hands than he could handle, the dispirited buckaroo traded the mare to a rodeo bucking string outfit. For several years, she was near-legendary among bareback riders. Finally, the "buckin' son of a bitch," as Tolbert affectionately refers to the catch of his life, was sold to the rodeo star, Jim Shoulders. Before the mare's days were over, Shoulders was riding her in the National Rodeo finals.

For many years, members of riding clubs from Salt Lake City, Provo, and Nephi came on weekends to drink, barbecue, and chase mustangs in the mountains around Delta. Their principal aim was pleasure—and a live trophy as proof of their riding and roping skills. Often not skillful enough to run down and rope a healthy adult, the riding club members just about wiped out the colt population. If they didn't take the young ones home, they left them to fend on their own. Which, if young enough, meant that they'd die or be easy prey for mountain lions. As Tolbert puts it: "Five or six of them riding clubs came down here for years, and it got so bad that I remember on one trip across the desert I saw nothing."

In the early 1960s, Tolbert took the riding club problem to the county commissioner. By the time he got through pleading his case, he got the Millard County Commissioner and others in adjacent Juab and Beaver counties to agree that mustangs could not be chased from January through August. These are the months that include the foaling season and give the colts enough time to mature so they can make it on their own if separated from their mothers. The county commissioners also determined that a mustang had to be at least a yearling before it could be captured legally. Tolbert contends that largely through his efforts the wild horse population rebounded. But this didn't mean that there still wasn't a fair bit of mustanging going on. "And the best thing of all about all this chasing we were doing," Tolbert says, "was that the horses were regularly harvested at private rather than government expense." To Sherm Tolbert, the riding clubs were a hell of a problem, "but the government not minding its own business is always a worse problem. And a darn costly one too."

Tolbert also got the county commissioners to consider charging $125 for a license for chasing mustangs. But the proposal went nowhere, because a few people who had been doing it for years objected. Tolbert feels that that was a gross mistake, that if the license fee idea had been accepted "there wouldn't have been a BLM solution. Then the government could've gotten a lot of money, and the people could've had fun too. It's just as good a sport as fishing or something like that. You have a deer hunt to harvest deer. Why not do the same with horses?"

Sherm Tolbert revealed another strong prejudice: he has no use whatsoever for what he refers to as "drugstore cowboys." "That's people that wanna be cowboys but never will be and never have been." He explains that drugstore cowboys are really city slickers who dress up in fancy cowboy outfits and hire airplanes to tire the horses and chase them onto low open ground. "Then," Tolbert says, his squash-yellow teeth jumping out from his long face, "they just wait there, sitting on their quarter horses all prettied up. Why hell, when the horses are all run out anybody can catch one. Even little old ladies who never been on a horse. Then those drugstore cowboys, they'd come into town and brag about how many they had caught. You couldn't pay any attention to them. You just had to laugh at them." He laughs, then asks if I've got enough time for all the stories he still has to tell me.

Spaniards brought great numbers of horses to the New World in the sixteenth century, first to the Caribbean and Mexico, and then to Florida. By 1675, there were horse ranches in Texas and New Mexico, and within another fifty years, in Oklahoma and Kansas. The horses were raised on the open range, and with plenty of feed and few competitors they multiplied rapidly. Many quickly became feral. The Spaniards had laws that prohibited Indians from riding horses, but before long they needed vaqueros for ranch work and had to ignore the prohibition. Within a short time, the Indians became good horsemen, began stealing horses, and then gathering them from wild herds. Horse culture among Indians rapidly spread northward from Texas and New Mexico. By the first quarter of the eighteenth century, Indians as far north as Saskatchewan, Washington, Oregon, and Idaho were mounted.

The Northern Shoshoni and the Nez Percé of Idaho and Oregon acquired horses early in the eighteenth century. When Lewis and Clark visited the Nez Percé in 1806, they spoke of "immense numbers of horses." They said they did not see "a single horse which could be deemed poor and many of them were as fat as seals." Three decades later another visitor to the Nez Percé wrote that he saw "large bands of Indian horses." He noted that "there are among them some very beautiful animals, but they are generally almost as wild as deer, seldom permitting an approach to within a hundred yards or more. No doubt some have never known the bit." Most of the horses owned by the Nez Percé roamed free much of the year. Some that belonged to this group, and to others such as the Shoshoni, may have escaped into the sparsely inhabited Owyhee country of southeastern Oregon and southwestern Idaho.

In the nineteenth century, cattle rustlers and horse thieves began passing their breathing booty along chains of rustling stations in southeastern Oregon. Occasionally, stolen saddle horses got loose and, difficult to recapture in the remote and inaccessible canyonlands, they became wild. Over time, their numbers were augmented by the workings of venturesome loners, horse ranchers, in a manner of speaking. They branded horses, let them loose on the open range, and then when the market was right, rounded up what they could.

The better stallions, like those in Nevada, Wyoming, and Utah, were sold for use in the Boer War and the Great War. Throughout the West, Italian, French, and English agents traveled about placing huge orders for cavalry stock, paying between two and five dollars a head. Some horses were sent to the Midwest to be used for light farm work, for hauling milk wagons, for racing, and as children's ponies. Well-broken horses always found a market, but unbroken mares and colts were difficult to sell. Mares were either returned to the range or, along with poorer quality stallions, sold for their hides, manes, and tails. There was a brief slack in demand after World War I, but it picked up in the 1920s when chicken feed producers and pet food canneries clamored for horsemeat.

Still, by the late 1920s, an estimated ten thousand wild horses were roaming the Owyhee desert of southeastern Oregon, far more than the niggardly moonscape could reasonably withstand. A business-minded sort by the name of Archie Meyers came into this country in 1928, figuring that everything that was unbranded and on the public range was his for the taking. Meyers and his cowboy teams of four or five men caught several thousand horses. Almost all of the mustangs were gathered using cow ponies, large traps, and chase routes that forced the horses into inescapable canyons. Meyers tried to use a fixed-wing plane for the job, but the effort failed. The broken rimrock country required an experienced and arrogant pilot.

The difficult Owyhee country had to wait until 1936 and the arrival of Floyd Hanson to see a one-engine pilot who could successfully maneuver the mustangs away from narrow canyons and steep scarps. With bravado and recklessness and a threatening siren that scared the bejesus out of the horses, Hanson, who described himself as an aerial horse wrangler, screamed into canyons and often came within mere feet of crashing into rock walls. Profit was his sole yardstick for measuring success. If the horses were selling for twelve or fifteen dollars a head, he wanted five or six for his efforts. Apparently, because of their meager worth, Hanson had little compunction about leaving colts behind to die when separated from their mothers.

Another notable mustanger of this period was Jim Bailey, who spent the first half of the 1930s in the employment of the Jordan Valley Grazing District. Bailey built elaborate traps out of brush, piled rocks, and wire. He also used a *parada*, a group of tame horses. The *parada* was run into a herd of mustangs to cause confusion. The lines of communication thus broken between a stallion and his harem, it was, in theory, easy for wranglers to push disoriented horses into a trap.

The 1930s had some mild winters that contributed to Owyhee horse numbers, but 1932 was not one of them. It was a hard season, and hundreds of mustangs in and on the fringes of the Oregon Owyhee, southern Idaho, and northern Nevada, starved to death or were victims of winterkill. Many horses tried to prolong their days by eating the manes and tails of kin and friend. In one canyon, practically the entire floor was covered with carcasses. It was an unforgettable disaster, yet a mere minor setback in the irrepressible imperative to reproduce. The Owyhee wild horse population continued to grow, and though mustangers whittled away at their numbers, the horses were increasingly seen as a problem by the Grazing Service, the federal agency created by the 1934 Taylor Grazing Act and the forerunner of the BLM. The Grazing Service ruled that cattle and sheep had first, if not sole, claim on the public's grasses.

The Taylor Grazing Act established grazing districts, as well as grazing access to public lands for a nominal fee. Unlike frontier days and the early decades of the twentieth century when the public domain could be raped at will, the Taylor Grazing Act precluded stockmen from running unlimited numbers of cattle and sheep on public land. The act restricted the number of livestock according to the carrying capacity of a particular range. Within this framework, horses, like cows and sheep, had to have permits to graze. Wild or unbranded horses were no longer welcome, and it was now the government's business to get rid of them—all in the name of better range management. At first, the government gave ranchers notice to clear off their horses. Some did, but others just caught as many as they could, corralled them until federal inspectors left, and then turned them loose. Frustrated, the government began putting out contracts to have the horses shot or eliminated in any way possible.

When the Second World War came along and food demands became more pressing, the fiery Harold Ickes, Secretary of the Interior, reaffirmed the government's stand against mustangs. In his words, "The removal of wild horses would protect the range in the interest of legitimate cattle and sheep production to win the war." The Grazing Service was directed to supply Russia, Italy, and France with tens of thousands of pounds of horsemeat a year, while the Department of the Interior, under the auspices of the United Nations Relief and Rehabilitation Administration, would coordinate efforts to meet the horseflesh demand in underfed countries where it was a regular part of the diet. The source of all this protein was to be the West's wild horses.

The government's design was aided by the opinions of ranchers who were fed up with mustangs fighting with their prized stallions and luring away gentle mares. With the increasing popularity of tractors, there was also less demand for heavy work stock. Ranchers turned to raising thoroughbreds, quarter horses, and American saddle horses, which, when crossed with heavier-boned stock, produced excellent cow horses. It was thought that on the open range these new breeds didn't have a sporting chance in competition with mustangs.

In Ickes' plan, the Owyhee Desert in southeast Oregon and southwest Idaho was singled out because of its excessive wild horse population. Covering more than four hundred fifty square miles, an area one and a half times the size of New York City, the rugged Owyhee was thought to have more than five thousand mustangs in the 1940s. As food demands increased during the war, Interior's wishes made their way into the courts. In early 1944, the Malheur County Court in the Jordan Valley of southeastern Oregon ordered ninety percent of an estimated three thousand mustangs east of the Owyhee River rounded up. They were to be sold at public auction. Malheur County would use the proceeds to build new roads.

According to the *Malheur Enterprise* of May 4, 1944, the roundup was to be an historic event, "one of the last wild horse gatherings in the West." One hundred buckaroos signed up for the drive. With romantic flair, journalists wrote of how the cowboys would be roaming an uncharted landscape of jackrabbits, rattlesnakes, and eagles, at night throwing out their bedrolls and saddle blankets under a pristine sky, and in the morning rigging up stoves from old hay-rake wheels. Their bellies full, they'd then gallop "right through rimrock, sagebrush and badger holes" after untamed horses all but impossible to catch. But what they would catch—the romantics believed—would surely be a veritable genetic rainbow: blue and strawberry roans, pintos, canellos, sevinas, grullas, duns, blacks, browns, chestnuts, albinos, sorrels, and "sometimes one of those rare ones that looks like a zebra." The luster of the memorable adventure, the journalists proclaimed, would not be dulled by savvy buckaroos who said that many of the horses "were uglier than the ugliest women on earth [with] wild matted manes as heavy as a bundle of straw on their necks and tails so long that they sweep the desert floor clean."

Just about everyone was peddling a theory on how best to catch the horses. Some said it was all a matter of knowing the country: the horse trails, the water holes, the windswept canyons, the mountain gaps. Others said that success depended on getting the wind in your face and the sun bright and low behind you, at which point you could get within fifty or seventy-five yards of the mustangs. At that point, roping them would be a cinch. Still another theory held that the most successful buckaroos would be those who rode a broken mustang. When the horse got near one of his slick brothers or sisters, he'd whinny and whistle and throw his head and tail high in the air. He'd know exactly how to get his rider into position for roping.

With the aid of an airplane, it is alleged that the buckaroos somehow managed to drive twelve to fifteen hundred horses south onto open, relatively flat desert country. For almost a dozen miles, all went well. It looked as though this might prove to be the historic roundup predicted. But then suddenly, the horses—now a great wave of jostling, anxious bone and swelling flesh—began turning, kicking, and fighting one another, looking for exits from their captors. The cowboys shouted madly, swung their hats and ropes, hit the horses on their heads and faces with quirts. Efforts were futile, breakaways multiplied. A few on the run became many on the charge—a herd, nay, something resembling a battalion in disarray. From above the mustangs looked like long, broad shafts of cloud and dust and legs and hinds and tails galloping for the horizon. And freedom. It was said that here and there the wild ones "charged like a whirlwind into the frightened saddle horses." Riders became anxious; they were thrown and they were hurt. In all, the pursuers came up just about empty-handed. When it was all over, the cowboys could do little more than reassure one another that five hundred of them couldn't have kept the uncivilized beasts in check. They concluded that trying to bring that many horses in at one time had been pure folly. A madman's dream.

Other efforts were made to round up the Jordan Valley mustangs. Cowboys worked the horses in manageable bunches. Small camps were set up in the breaks, and traps were built in draws and canyons. A few roads were put in to truck traps in and take horses out. Now the buckaroos could measure success by looking at what was in the corrals. But what they saw was nothing like what they had hoped for. In three months, a mere five hundred mustangs had been brought in.

The captured horses were sent to a livestock auction in Ontario, Oregon, where most found ready buyers. Fetching ten to sixty dollars a head, some went to cowboys and ranchers, others to children, the biggest and meanest and hardest kicking to rodeos. A few buckaroos bought them because they were curious about what they might be getting: special blood, a piece of a dying legend, perhaps a legendary cow-cutting horse. Not all of the mustangs were so lucky. The old ones, the crook-footed ones, the jug-headed ones, and the undersized studs wound up in cans.

The romance and the presumed cruelty of the roundups were not lost on Easterners, particularly eastern humane societies. A claim that most tickled the "ill-bred" Owyhee Breaks buckaroo was that the horses had been "driven over cliffs into the sea." The mustangs were almost four hundred miles from the Pacific!

The war ended and the Department of the Interior temporarily took its long snout out of the business of eliminating the West's wild horses. But this didn't mean an end to professional profiteering. Bill Johnson, a cowboy who normally bought and sold cows, came to the Owyhee country to make big bucks rounding up mustangs. Within a three- to-four-year period in the late 1940s, he allegedly corralled more than four thousand horses.

Bob Davis, who moved into the Owyhee in the 1940s and knew Johnson, wonders to this day whether Johnson really made any money for all the time and effort he put into the roundups. According to Davis, Johnson usually didn't catch more than a handful of horses in any one of his traps. Johnson had the expense of putting in roads and setting up traps and running down the horses and hauling them to market, there to discover that his catch wasn't worth much. A horse over eight hundred pounds was bringing about two cents a pound; those under eight hundred got half a cent less. Most of the mustangs caught by Bill Johnson were small. "A thousand pound horse was a pretty big one in the Owyhee in those days," Davis says, "and lots of times there would be an old stud who didn't weigh more than six and a half."

How does Bob Davis, himself a long-time mustanger of considerable credentials, account for Bill Johnson's roundup successes? The first thing Johnson did was hire an able pilot by the name of Lonnie Shirtliff. A crop duster by trade, Shirtliff had never chased horses before teaming up with Johnson. The challenge and the adventure of it all, however, proved to his liking. Shirtliff quickly learned essentials, and the one he thought most important was never to push the horses too fast. If they were moved too quickly, one bunch would lag behind and would eventually be lost. Shirtliff became so good that it was said that he could anticipate where the horses wanted to go before they had made up their own minds. After his successful tenure with Johnson, Shirtliff hired his own crew of buckaroos and horse-trap builders. He continued mustanging in Oregon and Nevada with his beloved J-4 Supercub well into the 1950s. Before he called it quits, he had, by some accounts, pushed more than nine thousand mustangs into traps.

Johnson equipped a main camp at Three Fingers Butte with forty-three saddle horses. This was surrounded by a number of smaller camps scattered throughout the breaks. Huge traps, some covering more than one hundred acres, were assembled in concealed spots—in canyons, around bends, just over a hill. Strips of canvas several hundred yards long were stretched on wire across principal passes through which horses might try to escape. When they saw the bleached white panels flapping in the wind, they were frightened and turned in the desired direction. Where was the ideal place to put a trap? According to Bob Davis, "It was any place where you could get horses inside a pen before they realized something was wrong. And another thing: the steeper the drop into the trap when they come over a ridge the better." Johnson preferred to do his most intense work in early spring. By then, winter had spent itself and yet the mares were still in heat. The studs were so preoccupied with fornicating or fighting one another that pushing them toward a trap was relatively easy. On the march to market, mustangs were controlled by putting a rope around the neck and tying it to the tail. This prevented the horse from raising its head and running. Another method was known as side-lining. Using a rope, a front foot was tied to a back foot with barely enough slack for the horse to stand straight.

A rancher himself since 1945, Bob Davis can easily tick off the names of other Owyhee ranchers who had livestock in the Owyhee when Johnson arrived in the area. Davis himself was running about a hundred cows at the time. In addition, there were another twenty-two hundred head competing with the horses for feed and water. And this doesn't take into account sheep, the number of which Davis now can't bring to mind. The biggest problem was a shortage of forage. "Why, with that many horses and sheep and cows out there, there just wasn't a whole lot left to eat in winter. Four or five years after Johnson took all those horses out, you could really notice the difference. Mule deer and antelope came back and soon you could see thirty-five to forty head in a meadow every night. We didn't know till then that the horses had been eating themselves and everybody else out of house and home."

In the 1950s, Bob Davis and four other ranchers in Malheur County began getting together in the fall to round up Owyhee mustangs on horseback. In a given year they might catch two to three hundred. "We kinda kept things under control so they didn't build up," the diminutive and retiring Davis says. "We could kinda break even on it expense-wise. Of course, you didn't get paid for your time. It was something you kinda enjoyed doin', and something that needed to be done. You just couldn't let 'em take over."

By the mid-1970s, several years after the Owyhee mustangs once again had had a chance to multiply without human interference, Bob Davis and several other ranchers began hiring out as part-time wranglers with the BLM office in Vale. Initially, Davis and friends, with BLM authority, did roundups on horseback. Then in 1976, the Vale and Burns BLM offices started using the more efficient helicopter. The BLM hired Dan Kettle, who had never seen a wild horse before his first chase. Kettle, however, had flown Hueys in Vietnam and, as a card-carrying forty-forty pilot—one who flew forty feet off the ground at forty miles an hour—he had plenty of experience getting in and out of tight spots. With Bob Davis beside him as teacher, Kettle soon got the hang of chasing mustangs. Whereas a year or two earlier Davis and the other buckaroos could only bring in three hundred horses a year, with the help of Kettle, this number more than tripled. In one day alone, Kettle brought in one hundred fifteen horses, forty-five on a single siren-screaming push.

To Kettle and the crew of wild horse specialists in the Oregon BLM, Bob Davis is the resident expert. "Hiring him was the best thing we ever did," says one BLMer. Among other things, Davis taught government cowboys the value of scouting mustangs on horseback and chasing them enough to discover their escape routes before going after them with Kettle's Bell 47. Davis also showed them how to build traps that required neither wings nor wranglers. As Kettle explains it, cheerfully and with admiration, "When we're out there chasing, Bob Davis doesn't like to see anything for them to see. No wings, no buckaroos, he doesn't like them to see anything until they're in the trap. It works. It really works!" No one that I met complained about Bob Davis's and Dan Kettle's treatment of the public's wild horses.

CHAPTER 2

THIEVES, GUNS, AND HORSEFLESH

Twenty-four miles west of Ely, Nevada, I turned north into Long Valley and eventually came to the end of the neglected tarmac and a divide in the road. Lying in sagebrush and greasewood nearby was a sunken-faced cow: prone, fat, and stiff as an iron railing.

If I took the road to the left, I'd come to the Warm Springs Ranch where Julian Goicoechea spent many a long, back-breaking day during roundups. Julian Goicoechea was one of three Nevadans indicted in a federal court in 1967 for illegally using an airplane to round up wild horses on public land.

I took the gravel road to the right and followed it for better than twenty miles before meeting a blue and white, low-hung sign that read "Blue Jay Ranch—Art, Sandy, Ted, Clay." On one side of the sign was a picture of a cow's head, and on the other side, the head of a horse. Below the horse's head, someone had scribbled MUSTANGS. Indicted with Goicoechea, Art Cook had supplied some of the trap panels and piloted the airplane that chased the horses.

I drove toward the Blue Jay Ranch, into a slight depression, past rusted water troughs and a windmill good only for scrap. Climbing slowly through a narrow gap and thin clumps of junipers and pinyons, into the lower reaches of the Maverick Springs Range, I soon came upon several haystacks in a clearing. I swung my pickup around them to my left on the lumpy dirt road, drove less than a hundred yards and there, just as the BLM had said, I came upon a poorly constructed corral. It looked right, out of sight and hidden below a rounded, balding hill. There was a short wing on one side.

I walked around the corral, trying to assess its location, guessing how it might be used. Then I went up to the top of the hill and looked south into Long Valley. A long, vacant view indeed. Ted Barber came to mind. He was the third person indicted with Goicoechea and Cook. He had ridden shotgun alongside Cook in the plane.

One-time coyote hunter, one-time owner of Ted's Flying Service in Winnemucca, one-time crop duster, logger of more than twenty-five thousand miles in gliders and airplanes, innovator in the 1930s of a rope and tin-can device for hazing mustangs in a plane, vendor of ultra-light Eagles in the 1980s, Ted Barber claims to have corralled more than ten thousand mustangs. He did it in Supercubs and Cessnas from 1933 to the late 1960s, primarily in Arizona, Oregon, and Nevada.

"You enjoyed chasing them?" I once said to him at his home in Orovada, Nevada.

"I did it for the money. Sometimes I went for a percentage of the horses, sometimes so much per hour. Those like me did better range management than the BLM. Would still do better." Cocksure, likeable, and as feisty in his seventy-plus years as the bucking bronc on his brass belt buckle, Ted Barber doesn't mince words when he talks about the inefficiencies of the BLM, the misguided ways of all forms of government.

On file in the Nevada State Historical Society is a letter that Barber wrote to Nevada's U.S. Congressman, Walter S. Baring, on January 24, 1960. In this letter, Barber strenuously objected to Baring's support of the 1959 federal law that made it illegal to chase mustangs on public lands with motorized vehicles. He was angry because the law more or less put him out of business. "There is nothing democratic or fair about that law," he wrote. Rambling on for four single-spaced pages, he told Baring that: (1) all horses on public lands belong to ranchers; (2) he knew ranchers who were shooting horses to get them off the range; (3) a lot of ranchers ran horses for which they didn't have licenses, and they often ran them on their neighbor's lands; (4) it was the BLM's responsibility to control the number of horses on the range; and (5) in his experience, using an airplane is the most humane way to catch mustangs. Barber added that if the congressman was worried about the humane treatment of the horses, then he ought to press for a law against chasing them on horseback.

Then, in a postscript, Ted Barber wrote, "I am no longer interested in running horses."

At the 1967 trial, the prosecution got a county sheriff and a state brand inspector, S. G. Robinson, to relate what they had seen in February of that year. Robinson said that he and the sheriff were driving on an old mountain road when they saw a plane diving, presumably working horses. "We moved on up over the summit and found a better vantage point, and from there, we were able to see seven head of horses coming off the mountain. The plane kept circling and diving sharply at the horses. During those dives and circles the plane would use a siren and we could hear a gun being used when the plane was at a low altitude. We also observed a rope about fifteen feet long hanging below the plane and some large objects tied to the end of the rope. We believed that the siren was suspended on this rope, but later found tin cans tied to the rope. The siren was mounted on the left side of the plane.

"After about forty-five minutes of this constant harassing and chasing," Robinson continued, "one yearling was unable to continue running and dropped back, while four other horses continued into the valley. The plane stayed with the colt, but far enough back so that the colt would not give up altogether and stop on them. After the plane worked the colt far enough down into the valley so that it could not get away, it left and went after the other four horses. We observed the plane harass and turn these horses at will."

In defense, Goicoechea said that he had gathered and disposed of all mustangs in the area in 1958, and that the following year he put out one hundred of his own horses. That meant, as he saw it, that from that point on all horses, whether branded or not, were his property. Since he and his business partners, Barber and Cook, were only trying to round up what belonged to him, they could hardly be charged with a crime.

The sheriff and the brand inspector verbally described the horses they had seen when the sheriff served Cook and Barber with summonses (Goicoechea was not at the roundup). They saw a black stallion—unbranded; a bay mare—unbranded; a bay filly—unbranded; a light bay filly—unbranded; and a steel-grey mare—branded. The branded mare was bleeding from several shotgun wounds. In court, Barber admitted shooting her from the plane to turn the horses.

The defense countered by producing a witness, a deputy brand inspector, who said that it is not possible to see the brand on a mustang in February, because the horses still have their thick winter coats on. To see a brand, he said, you would have to shear the animal. (This claim is questionable.) Subsequently, when Art Cook came to the witness stand, he said that after the mustangs had been taken to Goicoechea's ranch, and after the sheriff had photographed the horses and then left, they were sheared. All of them, he claimed, were branded. As evidence, Cook produced photographs of the sheared and branded horses. The jury found Barber, Cook, and Goicoechea innocent, and that pretty much finished what had come to be known as the Wild Horse Annie Law.

Today, Ted Barber says that he rounded up several hundred mustangs in his Supercub with Cook and Goicoechea before they were caught. By one reckoning, made by Wild Horse Annie after the trial, one hundred fifty-eight horses were shipped from Goicoechea's brand inspection district a month before the threesome were charged. The horses went to a packing plant in Newark, California. Barber says that they would have rounded up all three hundred to four hundred of the mustangs in Long Valley if they hadn't been caught.

When I talked with Julian Goicoechea in 1982, he was in his early seventies and superintendent of the ranch that he had owned at the time of the 1967 trial. He said, "I bred and ran horses as a sideline since the 1940s." He ran them along with several thousand sheep and cattle, on an outfit that was sixty miles long and thirty-five miles wide. By Goicoechea's recall, his biggest roundups had taken place in 1950, 1961, and 1966. Using two traps, and with the help of a single-wing plane and several buckaroos, he gathered more than four thousand horses in these years. Most of them were shipped to Yerington, Nevada, and Newark, California, where they were sold for fish meal and dog and cat food. Goicoechea guesses that over the years he gave as many as seventy or eighty of the better-looking mustang yearlings to kids around the state.

When I got to the Blue Jay Ranch, Ted Cook, one of the sons, about sixteen, greeted me. He was chewing a wad of tobacco. His father, he told me, wasn't home; he had taken some horses to a veterinarian in Elko. Ted said that he lived on the ranch with his brother and father, that they had about two hundred fifty cows, five saddle horses, and another twenty horses besides that. He didn't say what these other horses were used for.

"When that law came in my dad had a claim for one hundred thirty wild horses. They weren't branded horses, just claimed horses. Our horses. The BLM cleaned them out. I don't suppose there's more than fifty or sixty that live in these mountains now."

"What do you do with your extra horses?" I asked.

"We mostly sell 'em up in Twin Falls, but we ain't sold any in a long time."

The conversation turned to other matters, and at one point Ted said, "It's pretty nice living up here, but we get a lot of hassles from the BLM."

"About what?"

"*Everything*. They antagonize you about *every*thing."

I had to get Bill Lindsey who works for the BLM in Ely to tell me what might be included in Ted's idea of "everything." One complaint, it seems, involved a horse trap on the Blue Jay Ranch that the BLM tore down. Bureau cowboys also would have destroyed a loading chute if Art Cook hadn't hauled it away before the bureau got to it. Another problem was that in the summer of 1980, the BLM rounded up four hundred eighty-nine mustangs in Long Valley and the nearby mountains. Sixty of them had Cook's brand on them. Art Cook, no different than other ranchers, didn't—and doesn't—have a permit to have horses on public land. Not wanting to brashly make charges about how the mustangs got branded, or what Cook was doing with three water traps neatly concealed behind hills on private land adjacent to public land, the BLM gave him an opportunity to claim that the sixty horses were his. According to the BLM, Cook said he didn't want them. The bureau sold them, and then Art Cook protested that he had been wronged.

At the BLM office in Ely, I talked with Mark Barber, who has had dealings with Art Cook. I asked, "Does he have special permission to brand horses?"

"He doesn't have permission to brand any."

"Were the brands on the horses you caught in 1980 fresh?"

"We don't gather often enough to know."

"Has he sold any wild horses?"

"He would have to get them through a brand inspector, and the records show he hasn't sold any."

Sometime later, I asked Dave Goicoechea, Julian's outspoken nephew who works for the BLM in Reno, if Cook's traps might be used for rounding up cows.

"Probably not. In that kind of country, it's not hard to get your cows together, or get them out of pinyon and juniper. Also, you're not going to find a corral for cows down in a draw."

When I finally met Art Cook he was wearing a blue denim shirt pocked with cigarette burns, a dusty black cowboy hat tilted forward on his head, and he had a two-day growth of grey whiskers. He was sinewy, salty. I had been warned by several people in the BLM that he was "coyote," suspicious of everyone, and that if I wanted to talk to him about wild horses I had better do some fast talking and then be prepared to hightail it.

I asked him if he was running any horses. By BLM counts, there were then more than one thousand mustangs on Buck and Bald mountains and in the Maverick Springs Range, all of which border on or are part of his public grazing allotment and his handful of deeded acres.

"I got a few left," he said coolly. "The BLM took about five hundred out of here a couple of years ago and that thinned them out. I imagine there's two hundred or three hundred left in the Mavericks. We still got a claim out with the BLM. We never got all our horses when they had that claiming period after that horse act came in."

"Do you sell horses?"

"Every once in a while. We break horses and sell them." He sounded cautious.

"I've been told you've got some traps on your land. What are they used for?"

"Catching the cows."

"Why do you need traps?"

There was a long pause, then he said softly, "Well, it's just an easy way to handle your cows when you ain't got too much help. A man and a boy can go up there and sit around the water and catch what's coming in without runnin' the hell out of them."

"How many can you hold in those pens?"

"I think you better go see the BLM," he said, his voice rising. "They can tell you more than I can. Anyway, you're already pretty well informed."

Southeast Oregon experienced a severe drought in the summer of 1977. Water holes dried up and the skimpy desert forage was judged barely adequate for rabbits and field mice. Ranchers and environmentalists alike began insisting that the estimated three hundred wild horses on the Paisley Desert had to be removed lest they starve to death. A small number of cows regularly grazed the Paisley, but even these were removed when ranchers envisioned disaster.

In an attempt to save the horses, cowboys from the Lakeview BLM office built a water trap around a man-made lake pit known as Fire Lake. They constructed a corral one hundred to one hundred twenty-five feet on a side. Three of the sides were thirty feet from the water, while the fourth, where the horses would enter, was close to sixty feet from the water. The additional space would give cautious mustangs breathing room, a small sense of security once inside the trap. The trap had a sixteen-foot gate and was rigged to close by remote control. When it came time to put the Fire Lake water trap to use, other water holes in the Paisley Desert were encircled with wire and brightly

BLM water traps can often be operated from a distance with a remote-control device that closes the gate after wild horses have entered the enclosure.

colored ribbons. Beside each hole was a propane tank with a lighter. At regular intervals, the lighter flashed, sounding like a shotgun. Everyone in the BLM was certain that it would take audacious and desperately thirsty horses to venture into one of these water holes.

The mustangs fooled everybody. They came to the Fire Lake trap all right, but they took their time sizing up the situation and then, with few exceptions, left. They headed for the water holes guarded by the red and yellow ribbons and the ear-shattering blasts, broke through the wire, and merrily tanked up. Frustrated and unable to explain the horses' behavior, the BLM gave up. The hearty horses had won—and they survived.

During the next four years the Fire Lake water trap was all but forgotten by the BLM. That is, until the morning of June 10, 1981, when Ed Depaoli, BLM area manager for the high desert in southern Oregon, drove into the Paisley Desert to check on two areas where, by agreement with the BLM, ranchers were to have removed their cows. On toward noon, Depaoli came to the road leading to the Fire Lake water trap and, to his surprise, noticed fresh truck tracks. Unable to think of a good reason why anyone would go to this particular water hole, he drove in to see who was there. About a mile before reaching the trap, he came upon a horse trailer and two saddled and branded geldings. Nearby he found a couple of buckets of horse pellets and several containers that could be used for water. Becoming suspicious, Depaoli called the BLM to determine who owned the trailer. He then drove into the trap, where he found fifteen horses mulling outside the corral. Three of them were haltered; the others, he quickly discovered, were as wild as unjacketed lunatics. They whinnied and snorted, and then charged off into the desert.

The gate to the water trap was wired shut and inside were four bay mustangs: a stud, a mare, a filly, and a colt. The colt, about two weeks old, was trying to nurse both the mare and the filly. Its every advance was foiled; the reason was soon clear enough. There wasn't the slightest bit of evidence that the mare had recently foaled. Depaoli guessed that when the trap gate had been shut, mother and offspring were on opposite sides of the gate.

Depaoli also noticed that the trap's loading chute had been badly damaged. Someone had apparently been pushing horses into a stock truck, and some of them had kicked through one side of the chute. The battered side of the chute had been repaired with metal panels that didn't belong to the BLM. After inspecting tire tracks around the loading chute, Depaoli concluded that several loads of horses had probably been taken from the trap. He decided it was time to call the sheriff and the BLM and try to apprehend the culprit when he returned to pick up his day's catch.

Late that afternoon, a rancher and his son arrived in a stock truck. Immediately the rancher was read the Miranda rights and then asked what the hell he was doing catching wild horses on public land. Calmly, as if telling a gas station attendent to fill up his tank, he said that he had turned his own domestic horses loose to give them some much needed water and grass. Now he was merely trying to get them back. When asked about the unbranded horses in the trap, he replied, "Oh, I was just going to turn them loose." When he was asked why he'd brought a stock truck to the site, he went silent. Depaoli was certain that the man would soon be on his way to jail.

No charges were ever filed against the rancher.

"Why not?" I asked Depaoli.

"They felt they didn't have enough to go on." With a grin on his face, he then said that he knows another rancher who openly talks about a big gaping hole that he'd put in a BLM water trap while having a good time chasing corralled mustangs.

Thirty-five miles east of Fallon, Nevada, beyond the Naval Air Station and the sand dunes and a sprawling, ruby-red whorehouse, lies a bar and a gas pump and five motel rooms set amid a gathering of well-branched willows and cottonwoods. On a highway map, the place is known as Frenchman's Station. Not far from the old Pony Express Trail, it was founded in 1856 as a watering hole for desert-weary travelers. Today, Frenchman's unofficial population is forty-seven: a young married couple, their two kids, one mother-in-law, one dog, and forty-one cats. On a breezeless Sunday morning I stopped at Frenchman's to fill up on gas and get directions to the Clan Alpine Ranch.

"Do you know the Johnsons?" asked the slender woman behind the bar.

I said I didn't.

"Do you mind if I ask what you do?"

"I'm gathering information for a story on wild horses."

"Oh! *Now* I see why you want to see the Johnsons. I don't know if they'll talk to you about that, but I'll tell you how to get there. You just continue east until you get to a couple of trailers—that's Cold Springs. Then not far beyond there's a small gravel road on your left and you follow that in for five or six miles. You can't miss it . . . if you don't get lost." She laughed.

When Blaine Johnson greeted me on the steps of his mobile home he was wearing only blue jeans and a lemon-yellow visored cap. Glancing at his massive hirsute chest, I briefly wondered if he'd once been a tackle for the Detroit Lions. After I explained that I was doing a story on wild horses and several people had told me to look him up, he said, smiling, "Which people?"

"Ranchers," I said.

"You mean the BLM." He abruptly turned and began sizing up my pickup, staring at the out-of-state license plate. He peered in the truck bed like he was looking for something, then cocked his cap and scratched his chest. His rubicund cheeks and thick golden locks glistened. He seemed suspicious, undecipherable, then suddenly like a harmless teddy bear. In 1980, Blaine Johnson had become the first person in the nation to receive a federal conviction for stealing mustangs from public lands.

Blaine pointed to the fractured, green slopes of the Clan Alpine Mountains that rose abruptly behind his trailer. I tried to conjure up an image of Sir Walter Scott's Rob Roy in the foreground, a descendent of Alpin King of Scots for whom the Clan Alpines were probably named. "It's late now," he said, "but you can still see horses if you go down past the ranch about a mile and turn up into War Canyon. That's what you came for, isn't it? Look in the trees and on the higher ridges. There's plenty of horses up there. Up past the cattle guard you'll see a dead one alongside the road. I don't know how it got there. Must have been the winter. It couldn't have happened any other way."

I said that I'd heard that he had been convicted of stealing wild horses and I wanted to hear his side of the story.

"You should talk to my brother Dean," he said, laughing impishly. "I'm just the one who got caught. Yeah, it's too bad you missed him. He's gone to the rodeo for the day in Eureka. Dean's the one who really knows about trapping horses. He's caught a lot more than I ever did!" He laughed again.

Dean and his family live directly across from Blaine and his wife in a modest, single story house. The two brothers and a friend, who together had previously owned a cattle ranch near Grantsville, Utah, bought the Clan Alpine Ranch in 1977. As part of the deal, they got rights to water, eight hundred Herefords, and grazing privileges for three hundred seven thousand acres of public land. The Johnson brothers and their partner also found themselves with a whole lot of wild horse neighbors. Perhaps as many as three thousand.

Ever since the final decades of the nineteenth century, thousands of horses have roamed the Clan Alpine and nearby Desatoya Mountains and Ione and Dixie valleys. Some of the first ones may have descended from Paiute Indian ponies, others from those released by ranchers and by miners who couldn't find their El Dorado. By the early years of this century, there were so many feral horses in and around the Clan Alpines that a few cowboys were spending several months a year trying to catch them. One of the more famous horse runners of this period was a half-breed Indian who lived in a tent and went by the name of Coffee Pot, to some, Wild Horse Coffee Pot. Although he never owned a ranch, Coffee Pot had as many as five hundred branded horses and another thousand or so wild slicks to which he laid claim between 1905 and 1930.

Many of the Clan Alpine horses during Coffee Pot's era were sold for chicken feed at two to three cents a pound. Others were shot, boned on the spot, and then sold to a fish hatchery in nearby Smith Creek. Some went to feed trout as far away as Truckee, California. Occasionally the manes and tails were sold, though they usually didn't fetch much. More often than not the hides were left on the ground to rot. By 1913 horse buyers from San Francisco, Sacramento, and Stockton were coming into the Clan Alpines and surrounding valleys to buy horses for hauling coaches and fire wagons. Their demands were pretty specific: the horses had to be three to five years old, and they had to weigh better than a thousand pounds. Even then, if a horse wasn't thought to be worth at least a hundred dollars, which most of the mustangs weren't, it wasn't worth the time and expense of taking the animal over the Sierras.

Johnson explained that as far as he and his brother were concerned, there were far too many wild horses on their ranch. If they didn't get help pretty soon, he said, they'd find themselves out of business. The fact that the BLM claimed that it would like to reduce the Clan Alpine horse population down to two hundred fifty or three hundred was a lot of "bureaucratic bullshit" to Johnson's way of thinking.

We walked toward a small rise of cheatgrass and crested wheatgrass, dietary staples for both horses and cows. Looking toward Highway 50, barely visible in the distance, Blaine said that his brother Dean took care of the cows and he did most of the farming. They have a thousand acres in alfalfa that they feed to their cows part of the year. The Clan Alpine Ranch has what is known as a summer-fall allotment. From December to April the brothers run their cows in the valley bottoms; in the summer and fall, they're let loose in the Clan Alpines where they feed on Idaho fescue, bluegrass, and crested wheatgrass. Indian ricegrass, Johnson noted, was once common in the area, but because of the horses the inferior cheatgrass had largely taken over. Here and there he pointed to patches of cheatgrass and said, "That's our grass, we pay for it. The horses were helping themselves, so we started helping ourselves by taking some out. There's too many."

"How many did you and your brother catch before you got caught?"

He looked at me funny, then asked, "Were you in the war?"

When I said I hadn't been, he said "I was in for a while. I learned something from that experience. I found out free enterprise can do twice as much as government can do for the same money. You take the way the government fights wars. Inefficient, just plain inefficient. If they'd give those wars to free enterprise, it would be a whole lot cheaper for all of us. It's the same with rounding up and taking care of these horses. If the government would let free enterprise people like me handle the problem there wouldn't be a problem. The way the BLM takes care of wild horses reminds me of the Washington bureaucracy. All that inefficient bullshit they do."

Blaine wouldn't say how many horses he and Dean had trapped, nor would he give me the details of how he got caught. I was able to get most of the story, however, from George High, the Nevada BLM's principal enforcer of the federal anti-mustanging law. He played a major role in catching Johnson.

It seems that High was out patrolling one night in his camper north of the Clan Alpine Ranch when he saw a stock truck in the middle of the road and wondered what a rancher could possibly be doing that time of night. He didn't stop to find out, but he did jot down the truck's license plate number. The next day, following instinct, High went back to the spot where he'd seen the truck and there, along the road, he found a dead mare. The horse was clearly a mustang; it was unbranded and had an unkempt mane and tail. Only later would High learn that the mare had been chased just after she drank several gallons of water. Then, as fate would have it, she collapsed and died right in the middle of the road. When High happened upon the truck that fateful night, Blaine and a friend were trying to drag the corpse off the road. Blaine Johnson still wouldn't have gotten caught but for the fact that he and a friend tried to sell six mustangs and an orphaned colt five days later in Salt Lake City without a proper bill of sale. When the Nevada BLM was notified of the irregularity, High had no trouble making the right connection.

Blaine Johnson and his friend were charged in federal court with a "midnight roundup" of wild horses at a water trap and "the malicious death of a free-roaming horse." The friend was released for lack of evidence, but Johnson, at his lawyers' advice, pleaded guilty after initially claiming that the horses were his and had strayed from his ranch. He was given a three-year suspended sentence, a $1,000 fine, and required to spend two hundred hours cleaning out springs used by mustangs. Johnson's lawyers wanted him to plead guilty in order to challenge the federal wild horse law. They claimed that the definition of "wild free-roaming horses" is so vague that it violates the Fifth Amendment right to due process. The lawyers took issue with that provision of the law stating that "wild free-roaming horses are all unclaimed horses on public lands of the United States," arguing that "unclaimed horses" is meaningless since ranchers insist that all horses on public lands in the state are descended from domestic stock. By the end of 1984, Blaine Johnson's lawyers had gotten nowhere in their case against the government. On the other hand, Johnson had put in the two hundred hours he owed the public on behalf of the horses.

Blaine Johnson told me that he didn't think that stealing mustangs was much to get excited about. "I didn't get that many. And, anyway, what about all the others helping themselves to horses?" Johnson was referring to the fact that with only George High patrolling the entire state, and information on his whereabouts passed on faster than Hollywood gossip gallops down Wilshire Boulevard, it is all but impossible to catch Nevada's horse thieves. But even catching them with sweaty horses and a nearby stock truck is usually not good enough. Nevada's federal attorneys claim that they've got "higher priorities," and that illegal mustanging is almost impossible to prove. They won't take mustang cases unless there's "enough" horses involved, or—as George High says—"Unless you got thirty dead ones and the guy's standing over them with a white-hot gun."

George High might be accused of stretching the truth, but not by much. In 1984, two men pleaded guilty to stealing more than one hundred wild horses from Wyoming's Red Desert and selling them to a slaughterhouse in New Mexico. The federal district court judge in Cheyenne, Wyoming, made the thieves pay a total of $6,000 in restitution and put them on probation. They weren't sent to prison because, in the eyes of the judge, "they were following a long-time tradition where ranchers round up wild horses and sell them to slaughterhouses in order to make a living."

Before I left Johnson, he asked if I wanted to see the water trap where he and his brother had caught horses. It was, he said, about fifteen miles north of the ranch on the dry Edwards Creek lake bed. He gave me directions and drew a map. When he got through, he said, "You know, the government cared so little about that trap they hardly touched it after they caught me. You spend half a day fixing it up and you can catch all the horses you want."

Before I reached the road to the trap, I got curious about other dirt roads that gently slope into the planar vastness of the ancient lake bed. I took one that looked little used and was soon bumping and shaking my way over bitterbrush, sage, and deep ruts. After a mile or so I met a fence, got out, went through a gate, and followed an elusive trail for several hundred yards until I came upon a mushy swamp full of hoof prints and cloudy bowls of standing water. The water and the dark brown muck looked rancid and badly trampled. Gnats and flies were everywhere. Wondering whether the seep was being used by wild horses, I searched for their distinctive droppings. Nothing. Everywhere I saw nothing but cow paddies and swarming gnats.

As I walked back to the truck, I thought of a buckaroo's words directed at me a few days before. He had said that cows love nothing better than to stir up mud in a spring and, if the water's deep enough, "to get in about waist deep and do a back stroke." With a pretty good knowledge of both cows and wild horses, he was convinced that cows invariably do a lot more damage to water holes than horses, "because cows love to live in water." He said that whenever he'd seen mustangs around water, "they'd come in and stay long enough to tank up and then get right out. If there's many of them, they'll wait for the others and that's why it looks like horses spend hours around water. They might when it's real hot, but you're not going to see them make the mess cows do."

The illegal water trap was just where Blaine Johnson said it was. Crude, one hundred twenty-five feet in circumference, the trap was an imperfect circle of pinyon and juniper poles built around a well and water trough. The two dozen twisted poles, a good seven feet high, were sloppily strung with barbed wire. Around the well head was a rectangular cement base four feet high. At one end was a pipe that fed a twelve- by three-foot metal water trough. The long lips of the trough were bent and beaten, as if they'd been attacked repeatedly by a hammer-wielding drunk. The well and the trough were surrounded by milky pools of water and a healthy supply of horse scat. Some fresh, some old, some so ancient that it was indistinguishable from the parched earth. Several sections of the wire were draped with bristly horse hair. A crude wooden chute with a loading platform made from scrap lumber was connected to the north side of the trap; it looked marginally serviceable. The only entrance into the trap was on the east side. A single strand of barbed wire kept two angling poles from falling down. The entrance, better than a dozen yards from the water, was wide enough to allow a mare and a colt through at the same time. When used by the Johnson brothers, the entrance was rigged with several long, sturdy willows placed parallel to the ground, slanted inward, and interdigitated with one another. To get to water, the horses had to force their way through the "fingers," which were cut to sharp points on their inward-facing ends to discourage the horses from trying to escape.

Twenty-five yards southeast of the trap's entrance was a hole just big enough for a man to lie in a semi-fetal position. It was covered with a three-quarter inch piece of plywood. When the trap was first built, Blaine Johnson said he considered using an old mustanger tactic, that of putting a gate on the entrance and running a rope, covered with dirt, to the hole. Then, when a sufficient number of horses had filed in for a drink, he could close the gate behind them. Blaine had tried this method once and concluded that it wasn't to his liking. "I had better things to do than lay in a damn hole all night waiting for a bunch of horses to decide when they were going to get thirsty. Let the willow fingers do the work."

I walked inside and around the trap several times and couldn't help noticing the horses' large and numerous stud piles. An organic gardener's delight, tiny islands of mulch, I thought. Grass seeds pass through the alimentary canals of horses, and their droppings provide a great culture in which to grow. Sunflowers flourish in areas rich in horse droppings, and the alfalfa that one finds in Great Basin mountains has often been brought by horses.

Biologists tell us that up to ninety percent of wild horse defecations occur on stud piles, and the largest and greatest number of them are found along routes leading to water. This can add up awfully fast, since stallions and foals defecate twelve to thirteen times a day, mares about half this often. At the Johnson trap, the horses were in the habit of starting their scat piles about a hundred yards or so west of the trap. Closer to water, the horses use defecation, urination, and often fighting to remind one another who drinks first and who drinks last. Proud stallions with half a dozen or more mares in their harem don't want to lose their prize females. Stallions establish their superiority by defecating and urinating on the stud piles of subordinates. This is particularly true among the bachelors, who want to let other lonely males know who is next to take over for an ageing harem master. Displays of pawing, smelling, defecating, urinating—even bumping and pushing and smelling and biting one another around the neck, withers, flank, genitals, and rump—may last for a few minutes, or more than an hour. It depends on how desperate the males are.

One striking fact about stud piles is their enormity. Another is the variation in their ages, the fact that some piles are used for several months running. At the Johnson trap, the really big and important stud piles lie in front of the trap entrance. In all likelihood, this is where most challenges between the most kingly of the stallions occur. On my first visit, there was one distinctly fresh pile less than thirty feet from the entrance. Better than five feet in diameter and six- to eight-inches high, the age increased from the center outwards. There were several other stud piles to the east in a great arc around the entrance. Most of these were old and abandoned, a few showed signs of minor use by (perhaps) inferior stallions. One mighty old stud pile, some sixty feet from the entrance, measured better than fifteen feet across.

That night there was a three-quarter moon, and I wanted to see if I could get a bug's eye view of horses coming in to tank up. I decided to lie in the trap hole dug by the Johnsons, covered only by the piece of plywood they had left behind. Expecting a long night, a cramped one, and possibly a chilly one, I went to the hole with some crackers and a pint of my brother's prize 140-proof fire-copper moonshine.

The brittle blue of the evening was changing to ashen grey when I heard and felt the first small tremors of heavy feet on the march, coming from the direction of the naked-looking mountains. Then, almost as suddenly as the horses' approach had grabbed my attention, there was a profound silence. Several long moments passed before I heard birds chirping nervously around me. They stopped, as if on cue, as though the stage curtain had been opened and the winged creatures were ready for the heavyweight clamor that followed: deep, heavy rushes of air forced between dilated nostrils, tuneful nickers, snorting, more blowing, and then a deafening squeal that made my spine quiver. My curiosity begging for a front row seat, I pushed up the plyboard several inches and peered out. My eyes focused on the trap's wire and shaky poles. In my cluttered field of vision, darkened by the mountains in the background, I could just make out the outlines of three or four horses. They were motionless. Anxious, not wanting to miss what might come later, I dropped my head and gingerly lowered the board; and I tightened the cap on the bottle.

Several long minutes passed before I was again treated to thunderous noises, heavy-footed tap dancing, a thumping grunt, another squeal, feet repeatedly pounding the ground. A thoroughgoing stud fight? A rambunctious bachelor cruising the water hole, hoping to steal a good-looking lady?

Joel Berger, a professional biologist with the Smithsonian Institution who spent five years studying wild horses in the Granite Range of far western Nevada, found that the worst fights among studs have to do with females. He reported seeing one fight that lasted off and on for seventy-two hours. A fifteen-year-old was trying to protect his harem from an eager, sex-starved male less than half his age. They kicked and bit and chased one another, until the young stud finally won out. Berger also discovered that stallions, like other animals, are deeply concerned about whose genes they're caring for. When a stallion becomes the new master of a harem, he forces females less than six months pregnant to have intercourse with him. After ninety percent of these rapes, the pregnant mares spontaneously abort.

Finally, I had to know what was up, so I took another peek but saw nothing, or rather, nothing more than several bulky shadows mulling about. Smaller than I had thought they would be, not as svelte as my mind's eye had envisioned. More silence. Now I wondered if they were aware of my presence. Maybe they were. Maybe they had picked up a whiff of my body sweat, a chemical in the crackers, the sweet, pungent smell of the corn alcohol on my breath. Whatever, they still wanted to quench their thirst, for before long I could see a line of them moving closer to the trap, swinging out and around the loading chute. In the night that had now turned black and shadowy, the approaching horses reminded me of dignified mourners coming to a bier to pay final respects.

There were nine in the first bunch, including two colts. The mustang closest to me, on the far edge of the stud pile arena, some twenty yards away, was the biggest. The lord, I thought. And so it seemed. He pawed the earth. He defecated, smelled it, stepped over it, smelled it again. Now looking at the trap, now looking away, he kept his head high and alert. Several times he threw his tail about. Minutes passed, and still no one headed for the trap.

The first to make a move was a colt, followed by its mother, and then a smaller horse—a yearling, perhaps. The threesome walked through the gate, drank with apparent gusto, a head now and again rising from the trough, and then they filed back out and took a position somewhere to the rear of the rest of the family. Four more went in through the trap opening, drank—more leisurely than the others that preceded them—and then returned to join those who had already filled their bellies. Finally a stout horse with a long tail and an uncomely shaped back came near the stallion. The stud raised and lowered his head, looked to the rear, and then the two of them walked to water. When they finished, they exited the trap, the band lined up, the stallion took up the rear, and within minutes they moved off into the night and out of sight.

It wasn't long before another band, this one numbering eight, came to drink. It was followed by still another family of four. I could see this small band waiting some thirty yards away near the loading chute while the larger family drank. When the group of four had its turn, all of its members drank at the same time. No altruistic, protective stud in this household.

Another evening on another ranch, thinking I might be able to get a bird's-eye-view, I climbed to the top of a thirty-foot water tank. The imposing twenty thousand gallon cylinder stands to one side of a large corral. Nearby is a four- by eight-foot water trough just off the ground and, on the other side of the tank, a gasoline engine that pumps water from the well. Not wanting to scatter my brains should I doze or fall asleep while waiting, I tied down an arm and a leg to the top of the water tank ladder.

Like the night I had spent on the Clan Alpine Ranch, each band took its turn at the trough. No more than five or six imbibed at one time, and when everyone in a family had gotten their fill, the harem left as a tightly knit unit. On this summer night, the horses came later to drink, and the studs were more aggressive with one another. In one event, two studs faced off some fifteen yards in front of the trough, about fifteen feet apart. Each stood near a fresh stud pile. Silent, staring, once or twice pawing the scat and the dirt. Then the larger of the two—but not larger by much—sniffed the stud pile at his feet, stepped over it, urinated on it, and stuck his nose in it again. The other stallion more or less followed suit, but defecated rather than urinated. Now, their families quiet and watching from the sidelines, the challenging stallions cut small half circles around one another before moving in closer for scarring combat. They rushed together and bumped necks and banged at flanks. Then one of them lowered his fat neck and thrust it forward, abruptly backing away his foe. The opponent responded by baring his teeth and aiming for soft spots on the underside of the neck. The threat was enough to cause the first mustang to back up and throw his head forward. He lifted his forelegs off the ground several feet, dropped them, then let loose with a bawling squeal and a skyward shake of the head. Presently they were in close encounter: at killing distance. Both stood on their hind legs and did brief battle with their hooves. Touching, hammering, not touching. Backing up, heads lowered, forelegs bent and a foot or two off the ground, they paused—then again came together. High on their hind-legs, seemingly hugging cheek-to-cheek, they locked their forelegs together. Then they broke apart. The war having escalated more than perhaps either wanted, the two stallions slowly backed away, kicked their tails into irregular orbit, swung their heads from one side to the other. There followed a muted threat or two, first one bluffing, then the other, each with greater bravado. But no contact. The biggest stud defecated on a small pile in the center of the open-air theater. The other one, not acknowledging that he was smaller or less mighty, sniffed the scat, defecated, then sniffed again, longer and more thoroughly. Then he turned away and left to join his harem. Within minutes, the smaller of the two had his harem walking toward water. Last in the feral *equus* world often means first at the watering hole.

A little over a year after my first meeting with Blaine Johnson, I returned to the Clan Alpine Mountains to look for horses in Shoshone Meadows. Though I'd never talked about the meadows with Blaine, I'd been told by other ranchers and the BLM that mustangs were still heavy users of the lush grasses and natural springs there.

This time I approached the Clan Alpines from the east, shortly after sunrise. I had turned north off Highway 50 onto a cryptic dirt road that swings around the eastern edge of the Edwards Creek Valley. Within the first couple of miles, rolling over dozens of thick, fresh stud piles, I saw several bands of horses off to my right on the lower flanks of the New Pass Range. I tried to approach them on foot, but couldn't get closer than about two hundred yards. I was struck by the fact that all of the family bands had about seven or eight members, that some of them seemed utterly oblivious to nearby bands, no more aware of heavy hips and long backs or the value of uninhabited spaces than diehard New Yorkers. As I returned to my pickup, I recalled what Steve Pellegrini, a long-time student of wild horse behavior, had to say about crowding.

Pellegrini began studying wild horses in 1963 in the Wassuk Range west of Nevada's Walker Lake. At that time, there were eighty to one hundred mustangs in the range, distributed among twenty-three bands. Most of the families, averaging about four and a half members each, had one breeding-age stud; in some there were one or two subordinate males that had no access to females. These families maintained relatively discrete territories, they didn't go into areas claimed by others, and they found ways to get to water without crossing well-recognized property lines. When they met on trails, there were often confrontations and stud displays consisting of rearing, neck arching, and blowing. Some stallions fought and scarred one another; a few were badly injured.

Twenty years later, Pellegrini found that the Wassuk Range mustangs had multiplied three-fold, to two hundred ninety-one. The number of bands had increased to thirty-seven, and the size of the average family band had almost doubled, up to seven point nine. An increase in numbers and density brings on several significant changes in horse behavior. Now families wander all over the Wassuk Range, paying scant attention to territories or the presence of others. Now there is little fighting among studs from different families, and within a harem, two or three males have a good chance of passing on their genes. Now families often eat side-by-side, and they travel together for days at a time. Territoriality has been abandoned.

Somewhat like the lonely, the elderly, and the homeless in big cities, lone mustangs suffer the most in high density living spaces. While young males have enough energy to form bachelor bands and roam together until they are able to join a large family or feel bold enough and strong enough to challenge an ageing harem stallion, defeated studs have the worst of it. They're the first to die when there's not enough food to go around. Lone individuals occupy the marginal feeding sites, the worst observation posts, canyon bottoms between the home ranges of reproducing families.

Near the northeastern end of the Clan Alpines, the gravelly road leaves Edwards Creek Valley and slowly rises before following the perennially dry Shoshone Creek. Here and there, road and creek bed are hemmed in by spectacular red-brown escarpments, rock terraces, fisted-buttes, and high-walled canyons that beg to be explored, canyons that most certainly are home to stallions and their harems. When I finally bounced over a twisting rise, I didn't have to check my topographic map to be told that I was looking at Shoshone Meadows. At 4,800 feet, in what looks like a giant coffee cup, lies a thick carpet of hyphenated greens: olive, sage, blue, emerald, yellow. Beyond this bowl of plenty, the meadow fingers its way westward and upward until it peters out, until jagged rock, thin soil, small sage, and dwarf pinyon take over on the precipitous slopes.

I saw not a single horse in the meadow or anywhere on the imposing mountaintops. Still, I had no doubt that mustangs frequently ate here, had been here earlier that day. The size, quantity, and freshness of the stud piles were the measure of my imagination. Sitting on a high knoll that looks onto the meadow, I recalled my conversations with Eddie Allyn and Jim Brown.

If you happened to meet Eddie Allyn on the street without his moose-sized cowboy hat and pointed boots you could easily mistake him for a hardware salesman. He's small, unassuming, slightly sallow, now in the autumn of his life. But even if you didn't make this mistake, it's unlikely that you'd guess that he had once been a hired gun. Hired along with his friend Jim Brown to shoot wild horses.

Allyn and Brown only hired out once, and then only for a month. That was during the winter of 1946. The place was Shoshone Meadows. Allyn and Brown now talk openly and matter-of-factly about the massacre, even to strangers. From Allyn, I got the distinct impression that he does so because it happened so long ago, because then it was legal, and because he has a fixed notion of how quickly wild horses multiply and how they threaten the rancher's way of life. For years, Allyn has been a Nevada State Brand Inspector, which means that he has spent a good part of his adult life around ranchers and cows and horses. In Nevada, it is widely believed that the difference between a rancher and a brand inspector isn't worth talking about.

Allyn says he got involved in the Shoshone Meadows incident because Jim Brown's father had been in the business of supplying horsemeat to fisheries. Brown's father had heard that two large ranchers were fed up with mustangs eating grasses that they felt belonged to their cattle and sheep. Since the ranchers wanted their problem solved sooner rather than later, they let it be known that they were looking for a couple of sharpshooters. The pay: two dollars for a pair of wild horse ears.

"At that time there was a lot of horses out there," Allyn says evenly, "I mean, a *lot* of horses. Twenty-five and thirty in a bunch was nothing to see. You could sit up on a hill and as far as you could see, all you could see was horses. We killed sixteen hundred of them. We did it on foot for about a week, they was that thick. We shot some from horseback too. They was so thick we could get right up to them, within fifty to seventy-five yards. Of course, you had to get down wind. One took the mare and the other guy took the stud. If you laid them down, why the rest of them would just stand there and mill. We took out everything, the colts too. We kinda wiped 'em out of this one area, but there was a lot of horses in Antelope Valley and we never went in there. Where we was, nothing generally ever got away from us. We didn't even take the mane and tail. The hides weren't worth anything. All we had to do was cut off the ears and leave some hide between them so they would be easy to count."

Allyn used a 30–30, Jim Brown, a 30–40 Craig. "They was the best we had," Allyn says. "Lots of times we killed a hundred or more in a day. Our only expense was shells, and we didn't waste 'em. We might've wasted a few in the beginning until we learned. We shot for the heart; it was quick. If you wounded 'em or something, like shooting 'em in the neck, you didn't stop 'em."

Brown remembers a few crucial facts differently than Allyn. "We killed exactly eight hundred fifty-one that year," he says. "And there was still two hundred fifty watering at Shoshone Meadows when we left. We might have killed them too, but we got tired of it. Spring was coming on and it was time to work cattle."

Eddie Allyn's last words to me on the shooting: "You could say maybe it was cruel at the time, the way we shot them horses, but there was so many out there there was no grass for sheep and cows."

And those of Jim Brown: "We just done our job. It was like going to war. If we didn't do it, someone else would have. I didn't feel nothing, it was just a job. But what we killed wasn't that much when you think about it. In the 1920s, south of there in the Ione Valley, they killed two thousand two hundred horses. They had to."

Jim Brown might have added that during the Great War bounty hunters killed more than ten thousand mustangs in the Shoshone, Toiyabe, and Monitor ranges of central Nevada. At the request of cattlemen who had gotten fed up with the multiplying herds, the Nevada State legislature in 1913 passed a law that made it legal for "any resident of the State of Nevada . . . to kill any wild, unbranded horse, or burro of the age of twelve months or over found running at large on any of the public lands or ranges within the State of Nevada."

Some seventy miles north of Elko, the traveler encounters a whole list of suggestive place names: Wild Horse Mountain, Wild Horse Creek, Wild Horse Ranch, Wild Horse Reservoir, Wild Horse Recreation Area, and the town of Wild Horse. This is the southern portal of the Humboldt National Forest and, after spending weeks in the desert, I found the scenery welcome: jagged rock walls that shoot skyward from the road, mesalike vistas dotted with pines and junipers, and shrub-framed views of the trout-rich Owyhee River. There are also lots of plants around that horses are known to relish: sedges, rushes, fringed sageworth, lupine, wild onion. What better natural sanctuary for wild horses?

Hiking away from the road, I went looking for mustangs; to my surprise, I saw barely a baker's dozen. I saw little evidence of horse trails, and only a few insubstantial stud piles. Finally, I stopped at a roadside restaurant to ask the whereabouts of the reality behind the alluring place names.

By the manager I was told the following: "Thirty years ago this place used to be full of horses. That was until the BLM came in and put in fences and called it better land management. They started bringing in these punk kids who had never seen a cow or a horse or a desert in their lives. They all went to some fancy colleges and got fancy degrees and then came out here to use their college knowledge to put in these fences. This was to help us manage our lands better. Huh! Then some hard winters came along and the horses got trapped behind these fences and got stuck in the deep snow. They couldn't get out. It killed them; it killed them by the hundreds. I don't know how many that stupidity killed, but lots of them starved to death. You can still see what happened. Go up to the canyons where the fences are and you'll see piles of horse bones. Fences in this country? Christ! It takes eight years of college to do *bull*shit like that."

I would later come upon a couple of facts that bore on this experience. In Nevada BLM headquarters in Reno, there is a six-foot high map showing the location of the state's wild horses and burros. The bureau, it turns out, has no record of a single wild horse within forty miles of Wild Horse. History lends a bit of insight into this toponymic irony. An old-time, rheumy-eyed rancher in Winnemucca told me that some time during the 1930s in northeastern Nevada, more than eight thousand mustangs were shot from the ground, from on horseback, and from a plane to clear the range for more sheep and cattle. The bounty hunters were paid three dollars a head. The money was put up by the Nevada Cattleman's Association. At a later date, the government got so frustrated with its inability to catch some of the remaining horses that it allegedly pushed the fleeing critters off high bluffs. I've not been able to document the Winnemucca rancher's figure, and I suspect that it may be something of an exaggeration. Still, who knows? It seems that many times both ranchers and the government have played God with a gun because they thought that there were too many wild horses.

Stinking Water Creek is thirty-three air miles east of Burns, Oregon, in rugged, timbered terrain. In 1978, the Burns BLM office estimated that the Stinking Water resource area had two hundred forty wild horses, exactly three times the number deemed appropriate for the area's carrying capacity. Actually, the BLM had two figures to work with: eighty was the maximum number that the bureau thought should be left after drawoff, forty was the minimum and, to some minds, the ideal. The fewer remaining, the slower the increase in the growth of the population and therefore the greater the interval of time before another roundup would be necessary. A little extra effort in the short run would save taxpayers money in the long run.

In October of 1978, the BLM set up a trap near Stinking Water Creek and sent in a helicopter and a crew of wranglers to help with the roundup. The trap was on private land because, by consensus among ranchers and BLMers, the homestead site was thought to be the best of the alternatives. A principal advantage was the relative ease of getting a stock truck into the trap to load the horses. Because the approach to the trap afforded easy escape routes for mustangs on the run, several steel panels were placed along a line of junipers to keep the horses on the straight and narrow into the trap. On the other hand, and contrary to the experience of other BLM cowboys and mustangers, the panels were not disguised. Chad Bacon, the easygoing BLMer in charge of the roundup, thought it was best if the horses saw the panels at a distance. This, he believed, would prevent them from running into the steel wall. Still, so that the horses wouldn't hit the barrier, small colored flags were attached to the steel bars. Although BLM wranglers would be present at the roundup, they would not be used to chase the horses the last hundred yards or so into the trap, as was customary on many roundups. Instead, a Judas horse hidden behind a juniper, would be released when the mustangs got within seventy-five yards of the trap. The Judas horse would lead them right into a corral.

The roundup was scheduled to take ten days, beginning October 12, 1978. An examination of the log book suggests that the roundup began on time, finished within a tolerable time span, and was successful. October 12—thirty-five horses caught; October 13—a bad day, only three were brought in; October 14, 15, and 16—twenty-three, forty-eight, and twenty-two trapped on each of these days; October 17—the pilot ran out of allowable flying hours and a day of rest was declared; October 18—another thirteen were pushed into the trap; October 19—the pilot was asked to do some roundup work elsewhere; October 20—the helicopter broke down and required minor repairs; October 23 through 25—thirty-three were brought in. Total number captured for the two weeks: one hundred seventy-seven. This was seventeen more than required by the planners, who had finally decided that 160 horses would suffice.

Elsewhere in the BLM record books another story was told about the 1978 Stinking Water roundup. These records show that during that two-week October roundup, forty "difficult" mustangs were gunned down. How, many subsequently wondered, could such a senseless act have occurred?

Chad Bacon says, "The real problem was that the horses had been chased several times in the past. They had been spooked, and some of them were spoiled and afraid of fences. And then the helicopter pilot couldn't get as close to them as he wanted to. The juniper cover was too thick. One way was to push the horses toward the trap in small numbers. As few as three, seldom more than ten at a time. We didn't put two groups together because then they'd be antisocial and a lot harder to handle."

No matter, some of the horses refused to cooperate. Stallions or mares—those at the head of a pack—seemed to sense that something was wrong. They refused to follow the Judas horse, and as soon as they saw openings in the poorly placed panels, they bolted for freedom along Stinking Water Creek. More often than not, horses fast on the heels of their leaders followed suit. After this happened two or three times, the bureau wranglers came to the conclusion that the lead mustangs were "problem horses." They decided that if something wasn't done, the enlightened ones would obstruct the roundup efforts.

Rifles were passed out to BLM volunteers, who were then taken up in the helicopter; as soon as a lead horse broke from the others near the trap, the order was given to shoot. As Bacon explains, "We took care of the problem with a 30.06. You picked your spot and it was simple. Almost all of them were shot in the heart. If you shot in the head you could have a real mess on your hands and we didn't need that. The overwhelming majority of those shot were studs. We took out four or five, never more than ten in a day."

"Did you know how many were being killed?"

"Nobody really kept a tally. The question was whether it was legal to shoot from a helicopter, not whether it was legal to shoot horses. It was something we felt was necessary at the time to reach management level figures."

Still not satisfied that I understood why the horses had been such a problem when they had caught as many as their resource people said were necessary, I asked Bacon if he could provide further insight.

"The horses had probably been down there where the trap was and when they saw something strange they got spooked. I guess that's the reason. I don't know why the horses did it. Those horses are creatures of habit and they don't always do like they're supposed to."

"You had to shoot them?" I insisted.

"That was the only viable alternative. We had to dispose of them."

"Could you have put the trap elsewhere?"

"Yes, we could have done some alternative trap work."

Within days after the final shots were fired and the last of the one hundred seventy-seven horses was rounded up, twenty-nine of them were released into the very area out of which they had been taken. The reason: it was decided that too many had been taken out, that the twenty-three remaining horses constituted a number too low for a viable herd.

When the Stinking Water massacre finally came to light, there was a BLM investigation at the national level. But as far as I can tell, it was largely perfunctory. No one was fired and only a few administrators got letters of reprimand.

Curious as to how the incident looked after the fact, I asked people who were involved or knew about it.

Jim Hicks, the Elko-based helicopter pilot who ferried the gunners within deadly range: "Do you shoot horses?" he asked a BLM cowboy unconnected with the incident. When he heard the reply, "Only if they're crippled or break a leg," Hicks brought to mind those who had given him orders to get close to the horses and then got angrier than a viper under siege by a mongoose.

A BLM wild horse expert in California: "That was highly distasteful and I can't believe it happened. My cowboys couldn't stomach the idea. They wouldn't have done it."

A BLM wild horse expert in Nevada: "I don't care about blowing horses away, but it's against the law to shoot them from a helicopter. That district manager should have been fired and had his retirement taken away. If the people who are paid to uphold the law can't do it, they should be taking a hike down the road. That guy who was the district manager when Stinking Water took place was promoted and is now with the BLM in Washington. He got an official reprimand for that killing and hung it on his wall."

Chad Bacon: "It was a judgment call. I don't think that anybody involved in the process enjoyed it or felt good about it. But it sure was a real experience to have the undivided attention of six BLM departmental investigators!"

After an unguided tour of the BLM's spacious and attractive corrals west of Burns—then full of good-looking horses that the bureau was having a hard time finding homes for—I headed south on Highway 395, toward jumping-off points at Alkali Lake and Hogback Summit and Paisley, to see more horses in their arid habitats. Along the long, lonely road, realizing that I'd be passing the turnoff to Beatty Butte, I got to thinking about the irony of Chad Bacon's role in the Stinking Creek incident. Bacon's the man in charge of Oregon's "primitive" East Kiger herd, which now roams a ten thousand-acre expanse of high desert along the east side of Kiger Creek some sixty miles east of Burns. To those with an interest in Spanish ancestry and the primitive markings of Mongolian wild horses and Tarpans, the Beatty Butte or East Kiger horses, which numbered about fifty in 1984, are easily the most precious single mustang collection under BLM control. Originally, the Kiger mustangs lived near Beatty Butte, an intimidating, sagebrush-mottled hump that rises nearly 8,000 feet above sea level. The East Kiger region was chosen for the horses because, according to one BLMer, "It was an area that nicely fit their numbers, and Kiger's a remote area where we can pretty well keep an eye on them." A four-wheel drive or a saddle horse is needed to get in to see the mustangs. Hunters go there for deer, but otherwise, besides the horses, the inhabitants are a few antelope and lots of jackrabbits, coyotes, and warbling birds.

Even a parochial dandy would have little trouble picking out an East Kiger mustang. It's small, rarely weighing more than eight hundred pounds, and is often dun or grulla (mouse dun), though background colors include brown, buckskin, palomino, chestnut, grey, blue roan, and sorrel. These mustangs have broad backs with a black stripe down the center from the mane to the tail, and zebralike markings on the legs. Some are dark above the hoof, often to the knee or hock; this is similar to the dark legs of Mongolian wild horses. The ears are frequently dark brown or black on the outside, with light interiors. For those with an interest in horse lineages, there are other things to look for in the Kiger mustangs: a somewhat convex head, very small nostrils, hooves that are small and hard, and curly fetlocks. And, if one has x-ray vision, five rather than six vertebrae in the lumbar section of the spine.

Bill Phillips, who joined the BLM some twenty-five years ago and is now a range specialist with the bureau in Susanville, California, was among the first to recognize the distinctive markings when the horses were rounded up in 1974. Since then, Phillips has taken a personal interest in the Kiger mustangs. He has urged the BLM to add to the herd whenever possible (about one percent of the more than four thousand horses that have passed through the Burns corrals have had the supposedly primitive markings), and has spent a good bit of his free time researching their possible ancestry.

Phillips believes that the horses' striking stripes and uncommon colors link them to Spanish mustangs. He will explain, as he did for my sole benefit with slides and an inspired lecture, that there are two races of true wild horses left today, the Mongolian and the Tarpan. The Mongolian wild horse, also known as Przewalski's horse, is a small, stocky, dun-colored animal with a dark upright mane, dark tail, and dark dorsal stripe. The horse has a white muzzle, narrow white rings around the eyes, and no forelock. The Tarpan, possibly the first horse to be domesticated, is mouse-grey with a thick head, long ears, a dark border around the ears, a short frizzy mane, a dorsal stripe, and zebra lines on the legs.

From these two breeds, Phillips explains, others evolved, in particular, the Sorraia, or native horse of Spain. The Sorraia took on features of both Tarpan and Przewalski horses. Seldom more than twelve to thirteen hands, the Sorraia is usually dun, though it may be grey or palomino. It has a black stripe down the center of the back and zebralike markings on its legs. Its head is large, its neck long, its hindquarters underdeveloped. Some of the Spanish horses that were brought to the New World were not genetically simple Sorraias, but rather, ones that had been repeatedly crossed with Arab and Barb horses, breeds introduced during Spain's seven hundred-year Muslim occupation. Other horses brought to the Americas were closer to being pure Arabs and Barbs. Phillips believes that there are still a few representatives of all of these horses in the West. "It is entirely within reason that some of these animals have a direct line to feral Spanish horses. They are not pure forms, but they are pure enough to show coloration and general structure of their ancestors."

What could account for the occurrence of these traits in a tiny proportion of the West's wild horses? According to Anna Drake, an evolutionary biologist of some distinction, there are at least three possible explanations. One, perhaps the most obvious and the one favored by Bill Phillips, is that the Kiger mustangs are direct descendants of Spanish horses with similar markings. A second possibility is that in the process of domestication, human intervention acted to conceal these primitive traits, but not entirely eliminate the genes for them from horse populations. Concealment may occur by selection for traits with dominant characteristics. For example, just as genes for brown eyes conceal those for blue eyes in human populations, it may be that humans have preferentially bred horses with characteristics that are dominant to ancestral traits and therefore mask them. By this reasoning, when horses are freed of human control and left alone in the wild, it is likely that the more "natural" genes will increase in frequency and eventually reassert themselves. If so, it may be that all domesticated horse populations have a reservoir of primitive traits that could be expressed if the horses were set free. A third possibility suggested by Anna Drake is that one or more mutations occurred in the Kiger mustang population. Biologists have found that different genes have different mutation rates. Mutations that occur frequently have a better chance of increasing in a population. It is probable that traits—such as stripes—that have become well established in several species of equids have relatively high mutation rates. With this line of reasoning, mutations for leg stripes are likely to occur eventually in any horse population. Once they do occur, genes for leg stripes will increase if they are adaptive.

The first hypothesis, the one favored by Phillips, makes the Kiger mustangs genetically special, a rather pure form of wild horse. The other two hypotheses suggest that traits such as those displayed by the Kiger horses are likely to occur, sooner or later, in any horse population released from human control.

I asked Anna Drake which possibility was most likely to be true.

"Who knows? There's almost no evidence to judge" she responded with the caution typical of a scientist. But then she smiled slightly and added with a trace of hesitation, "But if I had to place a bet, I'd guess those ancestral genes are still out there in domestic breeds, just waiting for an opportunity to show themselves again."

There is, in theory, a way to settle the dispute over the true lineage of today's wild horse herds. Geneticists have found that variations in blood enzymes and proteins reflect underlying differences in genetic makeup. The serum constituents of Spanish Barbs and Arabs and different domestic breeds could be precisely determined and then matched with that found in wild horses. This kind of work on genetic variation could add some welcome answers to the ongoing debate.

I asked Phillips how often he's seen primitive markings in domestic horses.

"When I was a kid I used to see more of them kind of horses. You could get carloads of them in those days." He added that quarter horses and others often have dorsal stripes.

Which is a point that some in the BLM like to make. A. K. Majors, a BLM manager in Lakeview, Oregon, who has rounded up mustangs in the Beatty Butte region, says, "Yeah, those Kiger horses have some markings and interesting characteristics, but that doesn't necessarily mean they came from Spanish horses." Majors claims that lots of mustangs around Beatty Butte were the offspring of work and ranch horses turned loose from an outfit called the Spalding Ranch. He says, "Then you have to know that there were extensive roundups of horses around Beatty Butte in the early 1950s. It is well substantiated that there were only fifty head left in that total area then. After that, people gathered them up through the spring of 1971. Sometimes it was the Klamath Indians who came in to get horses for the all-Indian rodeo."

Detecting a bite in his voice, sensing that something bothered him about the mustangs, I said, "I gather you don't think much of the Kiger horses?"

"We were angry because we took the trouble to get the horses, then the Burns BLM turned them back out. We thought, those people just want to make a name for themselves. There is no way, in my opinion, guys like Phillips can really substantiate where these horses came from, even if they killed them and checked their backbones, because there's plenty of variance in horses. It's just a matter of opinion. And what the hell difference does it make? A wild horse has been defined by law. It's any unbranded, unclaimed horse running free on public land. Anything else became a moot question when they passed the 1971 act."

Milt Frei, the nattily dressed, self-styled iconoclast who runs the wild horse program for the BLM in Nevada, is even more skeptical about the origin of the Kiger horses. "Well, Christ, that population they came from was not full of Tarpans or whatever they're supposed to come from. It was just random animals that come out of random matings that came out of wild populations."

Later, a scowl on his face, Frei added that the BLM had written up Phillips and the Kiger horse story in one of its multicolored annual reports. "You get miles of publicity out of that. In my experience, management agencies like this one like to pick out these little show pieces, and they don't mean anything at all. They really sell them and tell the public what a good thing they're doing. I think that Kiger thing is beyond what the 1971 wild horse act ever had in mind. And, plus, it's developed by those people who want to farm the horses. They did what I was proposing, but they did it in reverse! They created a young, healthy, thriving population. They just want to play around."

Frei's observations aside, the BLM has, on the whole, eschewed the idea of making a special effort to leave paints, palominos, or pretty horses with primitive markings on the open range. The bureau has wanted to avoid the charge that it is managing the horses in any way not specifically spelled out in law. Its range and wild horse specialists claim—rightly so, I believe—that the overwhelming majority of wild horses in the West are descended from domestic stock released by ranchers.

To all of this, Phillips would respond variously, depending on his mood and the company. Among those he considers hostile to his interest in tracing the lineage of the Kiger horses, he might well say, "I don't pretend to be an expert on horses, but I've delved into it enough that nobody needs to bullshit me. Lots of people argue that there's none of that Spanish bloodline out there. But where in the hell do they think that the horses in the western United States come from? Hell, there was horses here by the thousands before these settlers came. The horses that the settlers used were mustangs that came out of New Mexico, were gathered up and went back to the missions and came out with the wagon trains."

To others, he will confess: "I think we got a little bit of Spanish blood out there in the wild horses, but it's damned limited. All you can say about these horses when you get right down to it is that here are horses who have markings like primitive horses."

"You don't see yourself as promoting the Kiger horses, then?" I said.

"I can't beat drums, because I'm in the bureau. It's not part of policy to promote certain kinds of horses. Still, why not hold on to the primitive markings? I have a great fear that people who don't care will come in and shoot 'em out."

When Phillips is asked what he sees as the future for the Kiger horses, he replies that he really doesn't know. Then, wanting to let you know that he's got a lot more on his mind, he says, "I'd like to see the BLM expand the preservation of these horses, without getting into wholesale production of them. I would like to see several herds of them in different states. This particular kind of horse is unique." In an unpublished paper that he wrote on the origin of the West's mustangs, Phillips added that the public is already paying for the management of wild horses. "It would appear that the public has a right to expect the preservation of unique types as part of this management."

Before I left Bill Phillips, aware that he had once had an office down the hall from Chad Bacon, I brought up the Stinking Creek shootings. I felt compelled to ask how he felt about Bacon having responsibility for the Kiger horses.

Hanging his hand over his silver and turquoise belt buckle, inscribed with his name and the bold letters BLM, he said, "In Burns we shot difficult horses ever since we started running them. Hey, when you're running horses, it's a dangerous occupation. We primarily took out studs that were cripples or ran by themselves and would cut in and out."

Then, he said, "Yeah, Chad Bacon. I made him promise when I left that he'd take care of the horses. I threatened him with his life." He smiled, then said, "Besides Chad, I'm not sure anybody in the Burns district gives a damn about those horses."

After roaming the West to learn what I could about wild horses, I stopped at Central Nebraska Packing in North Platte, Nebraska. It was a dismal, rain-soaked afternoon, made all the more unpleasant because I arrived just as a hundred-odd horses were being chuted for an appointment with a gun-toting executioner. I soon learned that these were a small fraction of those processed in North Platte during the course of a year: twenty-five thousand to thirty thousand. Of this total, perhaps as much as ninety percent are sent to Japan, Belgium, the Netherlands, and France for human consumption. The rest are ground up and used for pet food.

Slaughterhouse operators have long recognized several classes of horses. Young, healthy animals above nine hundred or one thousand pounds are turned into steaks and a variety of exotic cuts. They command premium prices. In recent years, meaty horses have brought as much as forty to fifty cents a pound. Horses with less flesh and a defect or two have sold for half this much, and are ground into pet food. The worst of the lot, those which are scrawny and appear diseased, aren't worth more than ten or fifteen cents a pound in the best of times. Their destiny: bone meal and protein for fertilizer and cattle feed.

Those who worked on the assembly line at Central Nebraska Packing, and managers who bought the doomed animals, adamantly denied that any of the horses were wild. "They're all domestic, all from local farms," one of the office chiefs declared. Another inquiry concerning the origin of the horses prompted the reply: "None are wild, not a one. They come from ranches and the like in Montana and South Dakota, I don't know where. We don't take in wild horses."

I'm not much of a judge of horses, but a number of those I saw that day in North Platte appeared to be strikingly similar to mustangs I've seen in government corrals and on numerous treks into Nevada's mountains. As I left, I thought about the tens of thousands of horses that have been slaughtered there and in similar plants in Oregon, Washington, and Texas. I tried to reconcile my own cultural prejudices with those of pet owners and foreigners who support these "killer plants."

Americans, on the whole, are repulsed at the thought of eating horsemeat. Few other animals, except dogs and cats, are so clearly taboo in our diets. The reasons don't have all that much to do with tenderness, tastiness, or health considerations, but rather with the humanlike qualities we ascribe to these animals. We love them, we kiss them, we talk to them, we dream about them, we give them human names, we see ourselves or others in their personalities. As children, through television and books—the Walter Farley series, Anna Sewell's *Black Beauty*, and many, many more—we learn that horses are free, mighty, loving; they are friends. In our childhood picture-book relationships with horses, we are, if ever so briefly, able to extricate ourselves from the lives of adults who don't understand us. When we grow up and discover good literature, we again encounter romantic, loving, sentimental images of horses. Three of America's Nobel Prize winners, Ernest Hemingway, John Steinbeck, and William Faulkner, treated horses as something special. Faulkner's, *As I Lay Dying*, gives one of the noblest versions of man and horse to be found in literature. His book, *The Unvanquished*, provides the romance of war hero on horseback. In one Civil War scene in *The Unvanquished*, Faulkner has "the whole rim of the world . . . full of horses running along the sky." The horses "float, hang suspended in a dimension without time." In Hemingway's, *For Whom the Bell Tolls*, the horse is an object of love, and when the story's animal dies, it dies nobly.

And from Shakespeare's *King Henry V* (Act 3, Scene 7), we get:

I will not change my horse with any that treads . . . .
When I bestride him, I soar, I am a hawk.
He trots the air; the earth sings when he touches it.
The basest horn of his hoof is more musical than the pipe of Hermes.
He's of the color of the nutmeg. And of the heat of the ginger.
He is pure air and fire, and the dull elements
Of earth and water never appear in him,
But only in patient stillness while his rider mounts him . . .
It is the prince of palfreys; his neigh is like
The bidding of a monarch, and his countenance
Enforces homage.

Television, billboard, and slick magazine ads reinforce these early impressions. In cigarette and beer and perfume and clothing ads, galloping, wind-blown horses represent independence, freedom, a lost and idyllic frontier, an open-spaced refuge from the urban miasma. Polo ads implant an additional message: horses come with riches and the good life. Get a horse and be like the rich, these ads proclaim. In all this there is a discordant irony: we love horses and we won't eat them, but we conveniently forget that horses are regularly fed to millions of dogs and cats.

Not all Americans have had a problem enjoying horsemeat. When the Lewis and Clark expedition could not find enough game to eat, they ate their horses, "with as good Stomachs as iver we did eat beef in the States." Mountain men and military personnel would also eat horses if short on food. Indeed, some of them craved horsemeat, and would prefer it to available fish and fowl. Colts were thought to be better than beef.

Some native Americans also took to horsemeat, as I was reminded by Gussie Williams, an octogenarian Paiute resident of the Pyramid Lake Indian Reservation who converses in her native tongue. Mrs. Williams remembers times in the early part of this century when the Paiute ran out of cached food and the men didn't have any luck hunting deer and antelope. Her brother would then rope mustangs in the mountains west and north of Pyramid Lake. The meat was prepared by stone-boiling or by roasting in pit ovens. Although it has been half a century since Mrs. Williams last ate horsemeat, she can't forget its sweet taste. As I learned from her translator, "Not once did her brother kill the tame ones." Anthropologists who lived among the Paiutes discovered that they had a reputation as horse thieves. Frequently, they ate what they stole, while their own horses had prestige value and were consumed only under real threat of starvation. The Pyramid Lake Indians no longer eat horsemeat—so far as I know. They probably don't give any more thought to doing so than do run-of-the-mill pale-faced Americans.

At one time, horses were eaten by a number of American Indian tribes. The Apaches relished horse and mule meat, made it into jerky, and for a while, sold it to unsuspecting prospectors as beef. The Hopi castrated horses and burros to make the meat fat and tender. On the other hand, the Crow Indians wouldn't have dreamed of eating their horses, or anyone else's. As the Plains Indians came to appreciate the considerable utility of horses for warfare and hunting and transportation, horsemeat was removed from their diets.

The Navajo collected horses as if they were many-colored beads. A drought in 1905 killed thousands of their mounts, an event that some saw as a blessing because of the deteriorating condition of the range. By the 1930s, there were more than forty-five thousand horses on the Navajo Indian reservation in Utah, Arizona, and New Mexico, a large proportion of which was wild. When to this number was added ninety thousand goats, five hundred fifty thousand sheep, and thousands of cattle, there were far too many mouths to feed on the available Indian lands. The result was one of the largest roundups in the history of the Southwest. In 1939, the Navajos gathered in more than ten thousand horses. For the huge sale the horses were branded with paint: "A" for an average saddle horse, "B" for a marginal one, and "C" for those of such poor conformation and size that they were destined for dog and cat food. The "typical sales horse" was described as "decrepit, bony, a half-starved creature, barely equal to the long trek from the sales corral to railroad shipping points." Literally hundreds died en route, unable to make the slow drive across barren, water-poor range.

In the early 1920s, chicken feed processors in Hayward and Petaluma, California, began buying railroad cars full of mustangs. In one case alone, more than a thousand of them were driven from Riddle, Idaho, to Elko, Nevada, there to be prodded into boxcars for their fateful trip. Many of the horses were bought for a cent a pound, half this amount for yearlings. Colts that ran alongside their mothers fetched nothing. The railroads aided the slaughter of wild horses by offering special "chicken-feed rates," which meant that horses could be stuffed in boxcars liked oiled sardines and didn't have to be fed and watered.

It was about this time that the Chappel Brothers of Rockford, Illinois, came forward with the idea that pet dogs and cats needed a more balanced diet. Protein, it was advertised, could make them as healthy as their countless admirers. What wasn't advertised was the source of all this protein—horses. Initially, most of Chappel Brothers' meat came from Wyoming and Montana. Montana alone may have had a couple of hundred thousand mustangs, so many, according to one claim, that they ate "enough range grass in a year to feed one million head of sheep." The Montana and Wyoming state legislatures helped the Chappel Brothers by authorizing county roundups of mustangs, whose only alleged virtue was that of breaking down worn-out fences. The success of this midwestern venture—at one point, more than five hundred horses a day were being slaughtered in the Chappel Brothers plant—brought still greater demand for horses, and it brought in competition. By 1928, four companies were filling cans with horseflesh. Three of them were on the West Coast, including the largest, the Ross Dog and Cat Food Company. Based in Los Alamitos, California, the company got horses from half a dozen western states, including thousands of Indian ponies from Utah, Arizona, and New Mexico.

By 1935, scores of companies in the United States were producing canned dog and cat food. Humane groups complained vociferously, but to little effect. People were caught in a dilemma. Americans in 1935 spent more than $40 million to feed their pets. Within a decade, the number of horses slaughtered had risen from one hundred fifty thousand pounds a year to a high of thirty million. This was obviously a tremendous toll on the West's wild horse population, and by the beginning of World War II, the rate at which horses were being killed had dropped precipitously, to less than ten million pounds a year. Many pet food companies that had depended on equine meat either went out of business or were forced to look for other animal byproducts to satisfy the nation's dogs and cats.

Not all wild horses taken off the range wound up in pet food bowls. Some were bought for South Africa's Boer War. The British were willing to pay up to forty dollars for five-year olds, if "full-mouthed." Demand was also substantial during the First World War, when only Russia was able to meet its needs for "cannon fodder." Orders for America's stock were placed by Greece, Romania, and Serbia, not to mention the Italian, French, Belgian, and English agents who traveled throughout the Midwest and West to personally examine what they were buying. Few of the West's mustangs were able to meet the strict requirements of the British, however, whose war horses had to be dark and of a solid color, between five and ten years old, and "quiet, well broken, with feet properly placed and shaped, and without vice." Nevertheless, England and France bought close to a million horses from the United States during the Great War.

In the 1920s and '30s, five million pounds a year of cured, chopped, and smoked horsemeat went to Germany, France, the Netherlands, and Scandinavian countries. Portland, Oregon, was heralded as the home of a new industry, one which made pickled, frozen, and smoked steaks from Oregon's estimated two hundred thousand mustangs. China, the Philippines, and Japan were getting boiled and salted horsemeat that was processed on the shores of Pyramid Lake.

In the 1930s and '40s many horses were sold to southern farmers to pull cotton harrows when cotton prices dropped and tractors were too expensive to buy. Elsewhere, well-broken studs found easy markets for use as cow ponies. Mares and colts were more difficult to sell and often went for chicken feed and dog and cat food. Horse bones were sent to Italy to make buttons. Hides and hair were sent to Germany for shoes and mattresses.

Today, the Japanese consume over two hundred million pounds of horsemeat a year. When they cannot locate enough through normal import channels, they send buyers into the Midwest. Farmers in Iowa and Nebraska tell of local auctions in which Japanese entrepreneurs bid outrageously high prices for worn-out farm stock. They ship their purchases to the West Coast, there to be slaughtered, frozen, and sent home. One favorite Japanese recipe is "sakura nabe," or grilled horsemeat. The meat is cut into thin strips, threaded onto soaked bamboo skewers, marinated, and then broiled on a hibachi. Another dish of note is "shabu-shabu," or Mongolian firepot. Into a boiling moat, one puts slices of horsemeat, mushrooms, bamboo shoots, spinach leaves, chrysanthemum leaves, and tofu. The result is purportedly delicious.

French import figures are not all that far behind those of the Japanese. If one uses a per-capita measure of who's consuming how much, then the French eat more than twice as much horsemeat as the Japanese. Though horsemeat is not heralded in France's gastronomic guidebooks, the French eat it in a variety of ways: as steak or roast, stewed or braised, hashed or raw, even, though infrequently, in sausage. One French recipe strikes a wayward note; it is called "bifteck de cheval," or "horse beefsteak." The horsemeat is seasoned with salt and pepper, sautéed in butter, and then served under or beside an egg or two fried in butter.

The Belgians, the Dutch, the Swiss, and the Swedes also have a healthy taste for equids. Belgians, in fact, have the distinction of being Europe's premier horsemeat lovers. They consume about seven pounds per person per year, not quite twice as much as the French. The Swiss are fond of "fondue bourguignonne," deep fried horsemeat. Swedes prefer their horsemeat thinly sliced and smoked. That it occupies more than a minor place in Swedish life is suggested by the fact that horsemeat consumption is greater than that for lamb and mutton combined. This might well fool an unwary American, since Swedish horsemeat goes under the name of hamburger. But leave it to Italians, with their love of sauces, to come up with their own unique mix, a combination of ground horsemeat and pork called "salsa alla bolognese." Stirred in with grated Romano cheese and tomatoes and garlic and onions, the sauce is served over pasta cooked al dente.

Strange though all these tastes may seem to Americans, people in scores of cultures have eaten horsemeat since prehistoric times. In the Paleolithic period, horses were depicted as game animals in cave paintings in Spain and France; one eminent painting in the famous Lascaux caves shows a pregnant mare adorned with a fecundity charm. In the Dordogne region of south-central France, archeologists have found ample evidence to suggest that twenty-five thousand years ago, the protein content of the human diet consisted primarily of horsemeat. Seven thousand years later, reindeer had replaced horse on the caveman menu, perhaps because of declining availability of horses, perhaps because of changes in hunting technology.

The ancient Greeks did not eat horsemeat, but this was largely because the few horses available were needed to pull chariots and farm carts. The pagan Romans, less interested in horsemanship than emblematic statues would suggest, preferred onager, or wild ass, on their banquet tables, a gastronomic extravagance that cost many times the price of a slave. Caesar and the great historian Tacitus noted that horses taken in war were sacrificed. After they were killed, butcher-priests prepared the viscera for burning on the altar. The equine repast of the gods not consumed by the sacred fire was deconsecrated and eaten at a public banquet in the temple. This religious custom was of great importance in the diet of the poor.

Horsemeat was commonplace in Roman Britain. It was not forbidden until the eighth century when Pope Gregory III demanded an end to the eating of horsemeat because it was eaten by Scandinavian peoples in honor of the god Odin; too, the Pope saw it as a filthy and execrable food. Christian teaching prevailed—more or less. As late as the eleventh century, monks were eating horsemeat and giving thanks in metrical grace to God for its good and sweet taste. British Christians in Yorkshire also apparently didn't heed the Pope's message; a present-day derogatory term for a Yorkshireman is "kicker" or "kicker-eater," one who eats horsemeat.

During the Napoleonic wars, the surgeon-in-chief of Napoleon's Grand Army, Baron Dominique Jean Larrey, was so impressed with the amount of meat on the battlefield in the form of dead horses that he ordered it grilled on calvary breastplates, and seasoned with gun powder. Germans were similarly struck during the middle years of World War I. One high government official advised that slain horses should be immediately skinned and packed in sacks saturated with a solution of permanganate of potash, which would supposedly keep the meat fresh for weeks. He went on to recommend that the meat from fallen war horses should be fed to the million and one-half Russian prisoners held by the Germans, "since they would gladly eat it. In any case, it would be better than the frozen Canadian or Argentine meat which is set before our countrymen in English prisons, and whose odor and condition rouse such complaint."

Necessity has a way of exiling the most obdurate taboo. Paris police forbade eating horsemeat, or *chevaline*, in 1739, 1762, 1780, and

1784, "to prevent the diseases which such meats cannot fail to induce." But the edicts were often ignored. It was said that during the Revolution most Parisians virtually lived on horseflesh or cooked horse's blood. Although eating horsemeat was against the law, by 1810 the noted physician, Parent-Duchatelet, wrote that "The consumption of horseflesh in Paris is considerable and of ancient date, and it may be viewed as having become a necessity." By 1870, shops specializing in horsemeat began to appear in France's larger cities. Fifteen years later, Paris had eighty licensed horse butchers and Parisians were consuming more than ten thousand horses a year. Horsemeat was believed to make better soup than beef; some thought it a cure for tuberculosis. In the 1930s, one in ten Parisians ate horsemeat as it cost little more than half the price of beef. As late as the 1950s and '60s, Paris had several horsemeat restaurants that were frequented by health-food faddists.

Germans, especially poor Germans, also took to horsemeat. In 1904, when Berlin had a population exceeding two million people, more than sixty shops were selling three million pounds of horseflesh. On a per-capita basis, Saxony, with a large working class population, had even greater demand. By this time, Germans of all classes had cultivated a taste for horse liver, which brought premium prices. Inferior meat went into strongly spiced sausages flavored with garlic.

The demand for horsemeat for human consumption in the United States hit an all-time high during World War II. The reason was simple: there were shortages of beef and pork. Horsemeat was selling for twenty cents a pound, beef for fifty-five. West Coast meat packers advertised that they got their supply from "young wild horses—brought off the range." Plants in the Midwest and East got theirs from farmers. Only the choicest parts of the hindquarters and the shoulders were sold for table use; the rest, about three-quarters of the carcass, went into dog food. By 1943, one supplier alone, Hill Packing Company of Topeka, Kansas, was shipping five- to fifty-pound boxes of frozen horsemeat to seventy-five American cities. Boston received three railroad carloads a month, ostensibly for dog food; those in the know knew better. Yet, for all this civilian business, half of Hill's frozen and canned horse hamburger went to the War Department for the army's "Dogs-for-War."

Of course, even during the war, horsemeat accounted for only a small fraction of annual meat production in the United States. Sellers attempted to overcome strong prejudices by noting that horses don't get tuberculosis, tapeworm, or trichina, and that the vitamin content of horsemeat is three to five percent higher than that of beef, lamb, or pork. Outlets such as Milwaukee's Man O' War Market, South Bend's Western Plains Horse Meat Market, and Oakland's Scottie's Pony Markets advertised aggressively. In newspaper ads, they proclaimed that horsemeat was great in stews and meatloafs and hamburgers, equally good as roast or steak, that "it's similar to beef, prepared like beef," and that "you'll cheer its qualities with Mm's and Ah's." The Harvard Faculty Club added horsemeat to its menu. The club's Secretary, Carleton S. Coon, a world-renowned anthropologist, was one of several intellectual highbrows who liked its taste. Among those who wouldn't go along with the claims or the exigencies of war was New York's mayor, Fiorello La Guardia. He forbade the sale of horsemeat in the city on moral grounds.

It wasn't long before Americans forgot the war's privations and demands. In 1952, many vented primitive indignation when told that four-and-a-half million pounds of horsemeat had gone into so-called hamburgers in Chicago. Few were amused by the joke, "Have some chili con filly"; meat markets had to go out of their way to advertise that their ground meat was one hundred percent beef. But interest in breaking the taboo persists. In 1982, Chevalean Foods of Hartford, Connecticut, began selling horsemeat to United States Navy commissaries in New England. After Senator John Melcher from Montana convinced the navy that feeding sailors horsemeat was a bad idea, Chevalean Foods took their horseburger and steak sandwich business to the streets of New York City. Chevalean Foods said it was anxious to overcome America's "blanket of irrationality."

How, one might ask, does horsemeat compare with beef? Or is it, on its own merits, worth trying? Most who have given the comparison between beef and horsemeat serious attention agree that horsemeat is sweeter. The sweet taste comes from the high glycogen content. Some studies have shown beef to have thirty percent more fat than horsemeat. A T-bone steak has several times as much fat as a similar cut of horse, and a third less protein. Horsemeat can be tough because a horse's muscular fibers are more tightly packed than a cow's. This, however, is an avoidable problem: in fairly young horses the fibers have not had a chance to tighten up, whereas in older ones the muscle tissue has begun to break down. Some connoisseurs maintain that the right horsemeat properly cooked is . . . well, it's as good as beef . . . it's different . . . it's better . . . you can't tell the difference. The difference, of course, wouldn't matter at all if one were numbered among the poor, especially Third World poor.

In the mid-1970s, a Reno resident by the name of J. H. Robertson circulated a mimeoed paper with the catchy title of "Horses, Humans, Hunger." Robertson began his pitch with the ecological principle that everything is related to everything else: animals and plants to humans, and humans to humans everywhere. Robertson also believed that people are basically opposed to cruelty, whether to animals or humans. "Surely," he wrote, "nothing suffers more than a human mother helpless to feed a starving child." The global and humanistic context defined, Robertson made the observation that the West's rangelands are either fully stocked or overstocked. The fragile ecological base simply cannot withstand more animals. A very high percentage of Nevada's rangelands, which support more than sixty percent of the nation's mustangs, are in fair, poor, or bad condition—and they are deteriorating yearly. Just to reverse the trend would require, by Robertson's conservative estimate, more than one hundred fifty million dollars. Robertson made another point, namely that wild horse populations are large and viable, and growing at a healthy rate. His conclusion: the horses must be culled in order to preserve the range and prevent starvation—a fate far worse than death at the hands of man. It was at this point that Robertson arrived at the heart of his proposal. He advocated that wild horses above existing numbers, or some other agreed-upon figure, be shot by trained marksmen under the supervision of licensed meat inspectors. The horsemeat would be dried and then distributed through a non-profit organization to the world's poor. By Robertson's calculation, each horse would yield an average of three hundred pounds of dried meat, or enough for one pound per week for one year for six people. If a wild horse population of sixty thousand is assumed, and it is further assumed that the annual population increase is eight percent (Robertson accepts a twenty percent increase, but this is unrealistic), and that only the increase is culled, then roughly 1.5 million pounds of horsemeat could be sent yearly to the world's needy.

A mustang leaps for freedom from the holding pen.

CHAPTER 3

"I WANT A MUSTANG"

I had just pulled off Interstate 80 in Truckee, California, when I saw a father and his plump daughter adjusting stirrups. Curious about the small size of their horses and the unusual coat on one of them, I stopped to talk. I pointed to the daughter's gelded stallion and said that I'd never seen a color quite like it before.

"How would you describe the horse's color?" I asked the girl.

"He's a mustang!" the father cracked. "A *real*, genuine mustang! Once he was wild as a hare."

Mary Ball lives not far from the fairgrounds in Rock Springs, Wyoming. When I found her she was inside a corral feeding oats to her registered palomino quarter horse and her latest adoptees— Flicka and Sweet Pea. Flicka is also a palomino, then three years old and already topping one thousand pounds. Affectionate and gentle and every bit as handsome as the registered stallion, you would never know that a mere two years before, Flicka was freely roaming the range south of Rock Springs near a place called Buffalo Springs. There, ranchers were long accustomed to putting out palomino stallions, whose offspring they would sell at premium prices.

While I watched Mary feed the horses, she recalled how scrawny Flicka had looked as a yearling. "She was all ribs and lots of cheekbone. She was several hands shorter then too. You wouldn't have recognized her." Fixing the picture in my mind, Mary then added that, well, she just happened to like Flicka a wee bit better than. . . . I never did get the quarter horse's name.

In a vague sort of way, I thought I understood how she felt. I too was taken by Flicka, so completely in fact that she'd received just about all my attention since arriving. That was a mistake, as Sweet Pea accusingly reminded me at one point with a sharp nip on the back of my leg. I jumped and Sweet Pea whinnied and shook her head sumptuously. Then she put her glittering muzzle close to my face. I scratched her on the forehead and she let me know that she wasn't really unfriendly and prickly, just hurt. I felt a twinge of guilt at having ignored her. She forced me to admit that, as a child, I'd been prejudiced by the golden glow of Roy Roger's Trigger, that most famous of all palominos. The truth is that Sweet Pea, a luminous sorrel yearling, was already as comely as Flicka.

After Mary and I talked for some time, both of us giving Sweet Pea her due, Mary took hold of the horse's mouth and showed me her good teeth. They were all there, all twenty-four of them. When mature she would have thirty-six, four less than a male with his tushes. To determine the age of younger horses one counts the number of permanent and milk teeth. For those between six and twelve, age is guessed by examining the indentations or cups in incisor teeth. With still older horses, it is the slant of the incisors that matters. Being good at any of this comes from experience, and Mary Ball seemed to have plenty of that. She pointed out that horses which have spent much of their lives in sandy or gritty areas have the teeth of old mountain men.

After Flicka and Sweet Pea were broken, they would be put in a pasture with two other Red Desert mustangs that Mary had adopted a few years before, all of them at twenty-five dollars apiece. By her own admission, Mary can't seem to get too many of these horses. But she has to keep her costs down and that means putting them out to pasture away from home. Still, she rides all of them every chance she gets. Mary Ball gives the impression that she goes out of her way just to be around her mustangs.

As I left Flicka and Sweet Pea and Mary Ball that day, I thought of an experience a couple of weeks before when I had taken the Red Desert turnoff on Interstate 80 east of Rock Springs. I had swung around a half circle and come to a dead end at a rundown Amoco station, an overstuffed country store, and the crumbling Red Desert Motel whose heyday had been the '40s and '50s. In the store, a woman was tending the cash register. Her husband was sitting near the window, watching the pumps. His crimpy face was buried in a pile of old newspapers. I introduced myself, we talked briefly about wild horses, then I asked, "Do you know anyone around here who's adopted a horse from the BLM?"

"Who'd want one?" he barked. "Those damn horses are no good." He looked through me, then peered down at his twisted hands. "So you wanna know if I know anybody with one of those horses, huh? I'll tell you, it costs this damn government five to ten thousand dollars to catch a single one. [Two hundred dollars is closer to the real figure.] They go round chasing them in helicopters until they die. It used to be okay around here until the government came in. Everything the government puts its hands on is fouled up. Good thing we got Reagan in office." He shook his head, then added, "Naw, I don't know nobody with a government horse. And, anyway, those horses ain't no good for nothing. Giving them to people is just another waste of my good money," he said as he got up to help a customer.

The Wild Horse and Burro Act of 1971 eliminated rancher and free-lance mustanging and, left to their own, the horses quickly got down to the business of making colts. Many horse herds grew at the rate of ten percent a year, some—it is alleged by the BLM—at better than fifteen percent a year. Whatever the real figure, which no doubt varies from one area to another, it soon became apparent that something would have to be done about the exploding numbers. The law specified that herds were to be "maintained at 1971 levels"; ranchers were beginning to complain that the horses were taking a disproportionate share of their grasses and water; and a few, in and outside the BLM, saw that unchecked propagation of mustangs would eventually spell disaster for horses, wildlife, and the range itself. In 1974, the BLM told its Oregon wranglers to round up mustangs and put them up for public adoption. The Nevada BLM received a similar directive the following year. Though the horses were hard to catch on horseback—the BLM would not be allowed to use helicopters until 1976—the adoption efforts were successful, so much so that in 1976 the agency instituted a national Adopt-A-Horse and Burro program. Anyone with a large backyard and a hankering for a four-footed part of our national heritage could get a mustang by filling out a government form and paying a small fee (twenty-five dollars until 1982; two hundred dollars from 1982 to early 1983; one hundred and twenty-five dollars after that). If one is to judge only by numbers, the first decade of the adoption program was a conspicuous success. By 1984, more than thirty-five thousand wild horses had found foster homes in forty-nine states, with Californians and Texans taking almost a third of the total.

I had just been introduced to Ladel Bonham, who handled the paperwork for mustang adoptions in the BLM office in Susanville, California, when she said, "You know, everybody who adopts a horse wants a pinto. Everybody, and we don't get that many in. And Appaloosas too. We've even had applications come in that say, 'I want a quarter horse mare with a quarter horse filly by its side.' You gotta laugh." She did so gustily.

"Or, 'I want an Arabian with black hooves,'" Bob Moody, her craggy-faced BLM boss, added. "One little boy wrote in and asked for a white horse with a yellow mane and tail. But of course he meant he wanted a yellow horse with a white mane and tail!"

Ladel said, "If people see a particular color they like, they want it, even if the horse is ten or twelve years old."

"And it's foolish," Moody said. "Just plain foolish. Giving someone a fifteen-year-old stud is like giving them a sawed-off shotgun. About all you can do with those old ones is turn them loose in a big field until they die." Moody threw up his hands and an addled look crossed his face.

Ladel emphasized that people won't listen to advice on what kind of horse to adopt. She said she does her darndest to dissuade them from taking geriatric cases—selected because the rump resembles a crowded palette or because a twinkle in the mustang's eyes brings to mind a pony ridden years before at the county fair. "I never want them to take an old one." Would-be humanitarians and horse lovers, Ladel explained, seldom realize that patience and experience and hard work, even a good pinch of luck, are required to break an eight-year-old stud or twelve-year-old mare. Angry and disillusioned, or just plain frustrated when the uncivilized beast destroys a corral or refuses a rider, naive adopters want to blame the BLM for their problems. Some insist on returning their horse. More than a few have even demanded that Ladel or Bob Moody or one of his cowboys drop whatever they're doing and hurry out to a suburban two acres to pick up the mustang that the adopter was warned about. "Well, if they want to bring it back then we have no choice but to take it," Ladel said. "But they don't get their money back."

"There's something else," Moody sighed. "A lot of it is that they just don't want any horse. They want a wild one because they wanna say they have a *wild* horse. It's like people who wanna go somewhere to say, 'I've been there.' It's a fad."

All highlights, sherbet cheeks, and a brimming quart of good cheer, Ladel Bonham didn't require questions from me to keep her talking about her job. She hurried on like a charging greyhound, trying her best to accentuate the positive. "If it were up to me, they'd all get colts. Every last one of them." She seemed to bathe in the pleasure of seeing a young person get the right one. I got one of several examples of her concern when she asked if I would like to hear about her "little boyfriend," Chris Oliver from San Luis Obispo. When I nodded, she reached into a drawer full of files. She handed me the first of his letters.

"Dear BLM," Chris began. "I've been waiting for a horse a long time now. Then I was talking to my fourth grade teacher and she had the address for you guys. Then I wrote to you and you sent me the papers to fill out. After that I was all happy that I was getting a horse. But then you said it is two hundred dollars now and I was just wondering if you would let me have it free in the beginning. That is how I had planned it. I was just wondering because I would be one of I don't know how many. All of my friends have been asking me about the horse." The letter was signed: "PLEASE. Your friend, Chris Oliver."

Ladel promptly answered Chris's plea for help. Several times she called his parents to tell them that horses were available, and that they could avoid the eight-fold increase in the adoption fee by picking up a mustang for Chris before January 1, 1982. After some delay, the Olivers decided to travel to Susanville on December 24th, but at the last moment they couldn't find a trailer. When Chris was told that there were only three colts left in the corrals, he asked Ladel to choose one for him. Strictly speaking, this was against the law since Chris' parents had not paid the adoption fee. But even if they had, the BLM has no provisions for putting mustangs on a lay-a-way plan. Nevertheless, Ladel made an exception. "Well, you gotta bend the rules a *little* bit," she said. "I guess it would have been my neck if we had gotten called on it. I hate to admit it, but I'm a sucker for little kids. Their letters can get to you."

To get around the bureaucratic legalese, Ladel paid the adoption fee and had one of her cowboy friends pick out a handsome colt for Chris. She sent him a Christmas card informing him of his good fortune. Chris and his parents picked up the horse, he loved it, and soon he and Ladel were writing to each other.

In one letter, Chris wrote, "I'm very sorry I've not written to you before this. My horse Destiny is doing fine. We have been tying him up and he does not mind it very much. We were going to get him gelded Saturday so he will calm down. I just can't wait for him to get older so I can ride him. Last night on a commercial or something I saw what you guys are doing to the horses, feeding them to the lions. I know you cannot help it because you don't have enough money to feed the horses. If I could I would take all of them because I just can't see them getting fed to the lions. I wrote to see if they would put you guys on the air and if they do maybe it will give you a boost. What I mean is that people haven't seen what you guys are doing. If they do they will feel sorry for the horses and buy at least one. My opinion is to lower the price or give them away free so they will not have to be fed to the lions, or bring them down this way farther. Well, I have to go now." Chris signed for himself, "+ Destiny," and in a P.S. asked Ladel to write soon—and send a picture of herself.

"You've given away some of your wild horses for lion feed?" I asked Bob Moody.

"Yes . . . no; it's not that simple." And then he went on to explain what had happened.

In early 1982, the BLM Adopt-A-Horse center in Palomino Valley in Nevada gave Allan Hinman permission to pick up twenty-six "hard-core, unadoptable" mustangs. He was to use them for big cat food in his private zoo, Tiger Haven II, a one hundred twenty-acre farm in the Pahrump Valley of southern Nevada. Deeply concerned for the welfare of endangered animals, Hinman and his wife house and care for several dozen cats: leopards, ocelots, bobcats, some rare Asian and African species. They get them from zoo closings and owners who discover that their adult pets are more dangerous than a Saturday-night special. Normally, the Hinmans buy their horsemeat from a slaughterhouse in Texas, or they make the rounds of livestock auctions looking for the best price on anything that flushes red.

When the BLM raised the adoption fee for wild horses from twenty-five to two hundred dollars in 1982, some of those with thoughts of owning a mustang decided that being part of a fad wasn't worth that much. The horses began piling up in government corrals. By law, the BLM had only forty-five days to get rid of a horse from the time of capture. After that it could be put down, but not sold. Congress, in their 1971 law, made no provision for giving away "unadoptables." On the other hand, there is no law against giving the animals to people like the Hinmans.

When the Las Vegas Humane Society got wind of the BLM's Pahrump gift, several of its members just about went berserk. In their society's newsletter they called BLMers "dimwits," "bureaucratic maniacs," "pitiful slobs," "colossal clods." One angry soul vowed that

God would get the BLM for what it was doing to "His beloved creatures." "You can bet your cowboy hats, boys, when your time comes your civil service job protections won't help you a bit." Another denizen of Las Vegas was more to the point: "Let's cut out the red tape and just feed the BLM to the lions."

The Humane Society of Southern Nevada wasted no time filing a lawsuit against the BLM to prevent the horse slaughter at Tiger Haven II. The national BLM director, Robert Burford, responded almost immediately with an edict: no wild horses were to be killed or donated to anyone for any purpose. Those that had been sent to the Hinmans were to be returned immediately to Palomino Valley.

Milt Frei, the BLM's head wild horse specialist in Nevada, approved the horse giveaway to the Hinmans. He did it at a time when, because of the forty-five day provision, the Nevada BLM was openly talking of plans to destroy more than a hundred mustangs. The mile-high air of Nevada's wild horse country was filled with threats from a wide spectrum of animal lovers. Even those who want wild horse numbers greatly reduced were upset. Paul Bottari, the Executive Director of the Nevada Cattleman's Association, said, "Hell, anybody in their right mind who was aware of the situation *wouldn't* have done it that way. You're just opening yourself up for litigation from the wild horse groups and giving us more trouble than we already got. If you want to get rid of horses, send them to a zoo in Sacramento. Any place out of state."

I asked Frei if he thought his decision to send the horses to Hinman had been a mistake. He kicked a boot up onto his cluttered desk and said, "Yeah . . . it might have been a little irrational on my part. I tend to think we knew what might happen but it really didn't register. But," he chuckled to himself "*maybe* we did." If Milt Frei had a chance to reconsider his decision, he claims he wouldn't do anything different. He feels that feeding horses to worms is poor use of valuable protein. At every opportunity he has tried to put his attitude into practice. In late 1981, he shipped twelve horses to Folsom, California, for feed in a city zoo. A couple of months later he sent another fifteen to another zoo.

Milt Frei is not alone in the BLM in his humane pragmatism. Bob Moody, the BLM man in charge of gathering wild horses in far northern California, also has had dealings with the Hinmans. He still gets excited when asked about the horses that Milt Frei sent to them. "*Hey*," he says, "I tried to send some down there too! I got cut off at the pass. The flack and media exposure were so much we backed out of it. When it hit the papers I decided I wasn't interested in the reaction they were getting in Palomino." He added that about the time the Pahrump deal was being finalized, he talked with Allan Hinman on the phone and agreed to send him eleven stallions, then being held in Bishop, California. "The horses down there were very ugly. People who saw them said, 'You know what, these don't even look like horses!'"

Moody feels that giving away wild horses to people like the Hinmans is better than throwing them in a pit. "That's objectionable to everybody I know. All Hinman was trying to do was get cheaper food for his exotics." Moody can recall how the Pahrump affair wound up on television, the cameras allegedly focused on a lion chomping away on a horse leg. "But it wasn't a horse leg," Moody shouts. "I don't know what it was, but I sure know that cameraman didn't know the difference between a horse leg and a baboon leg."

Once Bob Moody received a letter from the national director of the BLM asking what the Susanville district was going to do with their wild horses when the forty-five day leave-or-die period expired. Moody didn't reply. He also didn't reply when he received a directive from the same Washington office informing him that horses held beyond the legal period had to go to the grave. "We never responded to any of that. No *way* you're going to get me killing a horse like that. My cowboys wouldn't put up with it."

Dawn Lappin, among the nation's most active and knowledgeable voices on behalf of wild horses, has feelings that parallel those of Milt Frei and Bob Moody. Concerned as she is with the future of wild horses on western rangelands, Lappin sees no problem with sending unadoptable mustangs to zoos or to others who will use them for worthy purposes.

Ray Tucker had just flown in from Antlers, Oklahoma, via Dallas when I met him at the BLM office in Rock Springs. When he arrived, I'd already been in town several days and was something of a familiar face in the district office. In a manner of speaking, I had been given a desk and was examining records on my own. Between phone calls and cowboys trying to make arrangements for a roundup in the Red Desert the following week, Tucker had somehow been forgotten. Filling a swivel chair, he looked confused. He introduced himself and right away, before I had a chance to say that I was not with the BLM, he said that he was in the market for up to five hundred brood mares. Could I help him?

"I'm after anything big and healthy," he said. "Anything between the ages of five and twelve or fourteen would be best." And, he added in his inimitable way, the mares better not have lost their interest in fornication. Now was the time to be picking them out and sending them home to Oklahoma. There would still be enough time, he explained, to put some fat on them before winter.

Tucker had come to Wyoming because he had heard through his horse-buying friends that Rock Springs was the best government facility in the country for quality mustangs. The Red Desert horses were large and peppy and, from all that he'd heard, they had good conformation. They were not like the "broomtails" and "jugheads" that he had been assured he'd find in Nevada.

Two years before, Ray Tucker had taken one hundred forty-eight brood mares from the BLM. He was satisfied with the results. "They're crazy as loons when you first get 'em, but they're no problem if you work 'em and you got the setup. I go in the back of my pickup with hay and they gentle down right away. We just chum 'em to gentle 'em with feed. If you got a little somethin' they like that. Then the next time you see 'em, they be lookin' for you. Do it a couple of times the first week, then a couple more the next week." He said that he didn't have any trouble catching the horses either. All of his pastures are fenced and, "every one of 'em got a funnel that leads to a trap and the trap to a corral." Once Tucker has broken the horses he finds they can do just about everything required of a good cow horse. "They gotta do everything on my ranches, so I know they're good."

Later, he said, "You hear about that horse school at Texas Tech where they try to determine the I.Q. of horses?"

I said I hadn't.

"One of them tests involves a complicated system of latches the horses open to get food. The colts I put in that school win every time. The little ones from these BLM mustangs, they're not gonna do as well as those schooled horses, but they'll do well enough." There was modest pride in his voice. I wondered if it was justified. There is evidence that humans select for dumbness when domesticating animals.

The intelligence of mustang offspring gave Tucker an idea. Not long ago he discovered that he had no problem selling colts foaled by his registered quarter horses. "Fact is," he said, "I can sell colts before they're born. There's a market for good horses from Alabama to New Mexico. If I can sell my quarter horses, I sure can sell them good lookin' mustangs."

Ray Tucker has spent a lifetime raising and selling beef cattle. He had nearly six thousand head on three ranches in southeastern Oklahoma when we talked. Besides what he feeds in his pastures, he buys young ones at around one hundred fifty pounds, then beefs them up to six hundred or seven hundred before taking them to market. Though clearly successful, Tucker had lost confidence in the cattle market; he felt hog-tied and corralled by the Board of Trade. "Every day they've got to run them prices up and down," he said. "Now you take a fella who wants a horse. He don't want to cheap it; he wants a horse that got reputation. You see him a-comin' and pullin' a trailer and he's come across two states. You know he's comin' to buy himself a good horse. It's not that way with cattle. There's about ten people in the U.S. who'll buy five hundred or more cattle and everyone of them talk to one another every morning and they decide what they're gonna give for 'em. It's not that way with horses."

Tucker sees Houston as his primary market. "There's a lot of oil money around that town and they all like good horses. Everyone of them are kinda cowboys or farmers. We sell 'em over the telephone. Just the other day, three of 'em drove up and they bought some colts. My horses are honest and everybody knows it."

Tucker had come to Rock Springs with a plan all worked out. He would keep all the first generation fillies from his mustang brood mares and breed them with his registered studs. Then, depending on how they looked and performed, he'd sell the second or third generation offspring. After the adopted mares have foaled three or four times he'll sell them to the highest bidder. When we talked, he had already halter broken one batch of colts that came from mustangs. To his way of thinking, they looked almost as good as those he got from his quarter horses. "We sure do like 'em and people begin to ask about 'em when word gets out. If we can get anything for 'em we can get four hundred dollars for 'em. If we can't get that, we're not gonna get nothin'."

Over the years, Tucker had cleared his ranch of undesirable brambles and briers, but then he began having problems with weeds, rank grass, and prolific clover. The clover got so thick that he couldn't mow it. He gave up on chemicals a couple of years ago, "because the price went out of sight." Then he thought he could solve the rampant weed problem with a thousand Spanish, slick-haired goats, which run behind the cattle and browse. But to keep them where they do the most good he had to build a lot of expensive fences. Then he discovered he had a predator problem, so he bought a barnyard full of Great Pyrenees dogs. When raised with goats, they grow up thinking they're goats, too, and become excellent bodyguards. Tucker's 1982 goat population was around two thousand, and this was more than enough. "Mares will do the same things as scavengers," he claims. "I got grass two feet high which is too rank for cattle. But you sure can put these mares in there. We got plenty of bluestem, Bahia, and other grasses too. So I think I got it worked out. The horses and cattle eat the grass, the goats browse, and my sheep take care of the weeds."

Tucker, like others who have seen potential for mustangs, formed a cooperative, and when he came to Rock Springs he had a hundred-odd affidavits in hand. These would give him the power to pick up several hundred horses. The government allows each individual to adopt up to four horses a year. In the first years of the adoption program, a single individual could just about take as many mustangs as he wanted. This was a cheap way for breeders to get brood mares, and for adults to give children an uncommon gift. But this multiple adoption rule also brought in an ample number of raw-boned crooks and ranchers who only wanted to turn a quick profit by illegally selling the horses to a slaughterhouse. In Oregon and Texas, Oklahoma and Utah, some took as many as seventy-five to one hundred horses at a time.

One of the more infamous examples of abuse came to national attention in the summer of 1978. A number of outraged Missourians contacted U.S. Senator Thomas Eagleton, claiming that a herd of nearly one hundred adopted mustangs near Lincoln, Missouri, was being poorly treated. When the BLM and the University of Missouri Veterinary School and a humane society looked into the matter, they found four dead horses and another ninety-three on the brink of starvation. The surviving animals were kept in a field whose edible vegetation had been gnawed down into the dirt. The horses had turned to eating toxic weeds, the only plant life remaining in the pasture. Many of them wobbled as they walked and needed immediate veterinary attention.

Because of this and similar problems, and a charge that up to ninety percent of mustangs adopted in large lots were being sent to rendering plants, the BLM set up a task force to look into abuses. When its final report was filed, the task force said it was able to account for only sixty percent of multiple-adoption horses. These mustangs were judged to be in "fair to excellent condition."

Until the Public Rangelands Improvement Act of 1978, wild horses remained wards of the federal government for life. Now the law allows an individual to get free title on up to four horses that he has had for a year, if a veterinarian certifies that they are being treated humanely. Yet, a great many horses still have no formal adoption papers. The reasons: the government cannot locate the adopters; those who have broken the law do not want title to horses they no longer have; and, for many, getting title means that their horse is no longer wild. And if it's not wild, so the reasoning goes, it has little glamour.

In April of 1979, senators Thomas Eagleton from Missouri and Paul Laxault from Nevada conducted a congressional subcommittee hearing on the worthiness of the Adopt-A-Horse Program. The hearing was in response to claims that the program had become an administrative debacle, and because "the BLM's operation of the program had brought howls of protest from all across the country." While cases of abuse were brought to the attention of the committee, it was also clear that thousands of people who had adopted mustangs had good words for the program. One homey letter entered in the Congressional Record was from Elmo Dugan of Mill Valley, California.

"I HAVE AND HAVE HAD FOR OVER TWO YEARS TWO HORSES (MUSTANG MARES) WHICH I ACQUIRED FROM BLM ADOPTION PROGRAM IN NEVADA ON MARCH 9, 1977.

"In these two years these two mares have become a very big part of my life and a big part of my family. If something was to happen to either one of them I would feel a very great loss.

"It would be like losing a child, parent or an immediate member of the family. You may not understand this unless you have had a wild animal and cared for it, and made a real friend.

"When I first received the two mares they had not seen very many people, never rode in a horse trailer, they were under weight with their ribs showing, shaggy hair from having no grooming. The first day they rode in a trailer they had a halter put on them.

"They immediately started getting two regular feedings a day of alfalfa and oats, within a week their weight came up. Within three days I could walk up to them and put a lead rope on their halter and tie them to the fence long enough to curry them. They learn in a hurry that you are not going to hurt them and they respond to your caring.

"It does not take long before when they see you coming they run to the gate to meet you and give a warm welcome.

"These two horses at present run on a very big acreage where there is lots of green grass to eat plus hay and grain. They have regular checkups with the vet, receiving the necessary shots and worming. They are provided blankets to protect them from cold weather and a barn to go into when it rains.

"They are shoed once every sixty days, and are ridden a couple of times a week, both by my son, myself and family.

"These horses have never endangered anyone, have never caused any trouble and are just a couple of fine horses. . . .

"I FEEL THAT WERE YOU TO PUT A STOP TO THE ADOPTION PROGRAM IT WOULD DENY OTHERS THE PLEASURE I HAVE HAD ALONG WITH MY FAMILY'S AND FURTHER A FATAL BLOW TO OUR COUNTRY'S FEW REMAINING HERITAGES BECAUSE THAT IS WHAT THESE MUSTANGS REPRESENT.

"Please DO NOT stop the adoption program . . . let us give as many of these magnificent animals a happy, healthy home as possible. . . . It will further enrich our endangered Species and Heritage. . . ."

From time to time the BLM sends out questionnaires to a sample of those who have adopted a wild horse. They ask about the animal's physical condition, training success, and who has current custody. I looked at about fifty of these questionnaires in the BLM district office in Rock Springs, Wyoming.

From a young girl in Lincoln, Nebraska: "She has turned into a real pet. She gets her favorite grain and sugar cubes. She's getting spoiled. We ride her around the farm and she spooks at things. . . ."

A Rock Springs townie who took a stallion wrote: "I wore a raincoat on him yesterday and he shied a little bit."

A Colorado father who got two horses for his children replied: "My fourteen-year-old son is breaking one of them. My eleven-year-old daughter can ride the mare."

One questionnaire was returned with only a single question answered, that asking about the horse's physical condition. The response: "Excellent!"

A young boy, about the same age as Chris Oliver, had this to say: "My dad is especially happy with her. She talks to him anytime he's around. She gets spooked easy. She really got scared the other day when a helicopter went over the barn."

A letter from Pinedale, Wyoming, conjures up images of a mighty self-minded stallion: "The only problem with him is he can jump a ten-pole corral anytime!"

There was this terse reply: "Mine died. An umbilical hernia."

And another, not much different: "The stud you gave me died. It had too many worms."

Frequently, those who adopt a wild horse get a lot more than they bargained for. Horses catch colds, get bronchitis, laryngitis, pneumonia, pleurisy. They can be afflicted by many kinds of parasites: bot flies and blowflies, lice, mites, ticks, pinworms, stomach worms, ringworms, roundworms, screwworms, and strongyles—the larger of which are known as redworms or bloodworms.

Although worming is mandatory before mustangs are released to adopters, worms may be migrating and escape the effects of the killing white paste. Recurrences of the determined little critters begin within a couple of weeks, and effective treatment can require up to two or three additional visits by a veterinarian. For enthusiasts who have invested six- or seven-hundred dollars in a registered quarter horse, a vet's fees are of modest concern. Not so for someone who only paid a pittance for their horse. Some mustang adopters feel ripped off when a vet has to be called and his bill exceeds the original investment. Some take a so-what attitude and hope that the worms will seek other homes. Often their horses die.

The BLM checks mustangs for equine infectious anemia, a blood-borne viral disease transmitted by blood-sucking insects that causes disintegration of red blood cells. Lactating mares pass the disease to foals, who may die from it. The horses also receive vaccinations against several strains of sleeping sickness, influenza, tetanus, and rhinopneumonitis. Horses are particularly susceptible to distemper, influenza, and rhinopneumonitis, a highly contagious respiratory disease of young horses that leads to pneumonia and abortion. Distemper, or "strangles," as it is sometimes called, is caused by a streptococcus bacterium. Transmitted by ingestion or inhalation of infected discharges, it is highly contagious. The bacteria can live outside a horse's body for as long as six months, just waiting for the right opportunity to invade and reproduce.

The BLM requires that adopters have a sturdy six-foot-high corral made of wood or pipe, and a shelter. Until it is relatively tame, there is always the possibility that a mustang will hurt itself on barbed wire. Mustangs have been known to leap over a seven-foot barrier. Corrals are supposed to measure eight or nine hundred square feet. The horses need the room for exercise and to avoid medical problems, particularly when kept together. Horses sometimes nibble on another's droppings, and in the process, pick up tetanus and blood parasites. Unlike domestic horses, young mustangs seem particularly susceptible to these microscopic animals. In the form of larvae, blood parasites burrow through arteries, cut off the blood supply, and eventually kill the horse. No one knows exactly why mustangs are more prone to problems with blood parasites than domestic horses; it may be because of their low densities in the wild. Continually on the move, they are seldom exposed to these parasites and therefore don't get a chance to develop resistence.

Wild horses have different eating habits than their domestic counterparts, and they must be slowly acclimated to their new diets. One sad tale I heard in Oregon concerned a teenage boy who adopted a mare and a colt. He had every intention of giving them the best possible care. Unfortunately, no one told him that you can't winter-feed wild horses in their first year in captivity with chaff—the leftovers from an alfalfa harvest. There is simply not enough protein in it. Chaff will suffice for a horse that has been getting good pasture grasses for the better part of a year, but mustangs aren't that lucky. Many live in niggardly environments, and they are continually competing with others for the same resources. Nature has not given them excess reserves to call on. Thus, any old thing that looks and smells and feels like good feed just won't do.

For the young Oregonian, these kinds of facts came too late. His mare and colt starved to death. For its part, the BLM could have done more. It could have given the young adopter an extensive list of do's and don'ts. The agency might have told the boy that you can't get by cheaply with any horse, that his mare and colt would eat twenty to thirty pounds of hay a day, and in a short while, this would certainly add up to a whole lot more than what he originally paid for the mother and son.

From June 1977 to March 1978, just over two thousand Nevada mustangs were rounded up and taken to government corrals at Palomino Valley, twenty miles north of Reno. Almost from the moment they were brought in, nothing seemed to go right for the horses. Lines of people were expected to show up to adopt them, but the weather turned bad and few came. None of the BLM cowboys at Palomino Valley gave much thought to overcrowding in the corrals, the fact that the animals could be held haunch to haunch for a couple of weeks but not for several months. As winter arrived and the horses put on their long-haired coats, the few people who did think they might like to adopt a mustang found the horses unattractive. There was an unusual amount of rain and cold weather in Palomino Valley that winter and this led to a further problem. Hay strewn on the ground got mixed in with sand, soil, and feces. The corrals had once been a feed lot, and when they were bought by the BLM no one cleared out the knee-high manure. On some nights, the muck froze and the animals could scarcely move. Furthermore, the pens had no drainage, and a low section of ground where many horses were kept was periodically flooded.

What ensued the BLM would rather forget. One hundred forty-six mustangs died and another ninety-eight had to be destroyed. Many contracted viral infections and pneumonia. Many more died from internal blockage. The sand and soil in the hay compacted, and once inside the animal's body, acids and body fluids had not one chance in a thousand of moving the adobe-like feed. The BLM tried to blame Mother Nature, and reasons were sought to justify having scattered the hay on the ground. Scattered hay, it was argued, would reduce the transmission of viral diseases, cut down on fighting, and give smaller and more timid horses a chance to feed.

Dawn Lappin, who worked at Palomino Valley at the time feeding and watering and adopting out horses, claims that there were other problems that cannot be easily explained away. "There were two young punks out there who were beating the horses. They had a fear of wild animals and they never should have been there. Everything was smooth as long as I was around, but as soon as I would leave they would bring out the cattle prods and stuff like that. But the last straw was when they did the autopsies on those horses. I went out to see what they had done and there were intestines and horse parts strung from hell to breakfast. Half a colt here, half a colt there, you wouldn't believe what I had to step over!"

The Humane Society of the United States and the American Horse Protection Association were quick to complain. In a joint lawsuit filed against the BLM, they said that the Palomino adoption facilities were operated in a "cruel, brutal and inhumane fashion, thereby causing the horses extreme suffering, injury and death." HSUS and AHPA were incensed that the horses were buried in mass graves. Lorne Greene of Bonanza fame and vice-president of AHPA said, "Anyone who sees the death ditches dug by the BLM and filled with hundreds of carcasses of wild horses will be angered at what has become our own holocaust in Nevada." Hope Ryden, an active wild horse protectionist, was a witness for the prosecution. She was particularly upset because some of the ill animals had been trucked to the burial site and then shot in front of other horses. Can you imagine the psychological effect on the horses as they watched their own kind being shot? Ryden asked.

Bill Stewart, who was in charge of Palomino Valley at the time, said that the problem caused by the unusual rains was quickly recognized and everything possible was done for the horses. But not until after the dieoff and the executions. Then, the Young Adult Conservation Corps was brought in. Its members built above-ground feeding bins and separate corrals for mares, stallions, and sick horses. The total corral space was doubled. Horses were moved to higher ground, and thirty-two hundred cubic yards of manure were shoveled and offered to the public without charge. Further gatherings and shipments of mustangs into Palomino Valley were halted.

Dave Goicoechea, who works for the BLM in Reno, claims that the Palomino Valley calamity was "an attitude problem." He says, "Those who ran Palomino Valley at the time saw a big difference between a riding horse and a mustang. To them rounding up mustangs was kind of like collecting coyotes and throwing them in a compound for a while. Some of those guys had shot horses before and thought nothing of it. They were only giving the mustangs as good feed as they figured they needed. They didn't see the Palomino Valley corrals as needing quality. They saw nothing wrong with throwing the hay on the ground and letting it get mixed in with all that stuff. It was ironic when those rains came along, because at the time there was no shortage of money. Those were the fat days when we shouldn't have had problems."

They often didn't—if they weren't caught. John Boyles, the national director of the wild horse program in Washington, says that in the early years of the program, ten to twelve percent of all horses caught were destroyed. "It was convenient because they were considered unadoptable." By the late 1970s, and especially after the 1978 disaster at Palomino Valley, attitudes within the bureau toward the horses changed. By Boyles' estimate, the percentage of mustangs that "had to be destroyed" dropped to less than two percent.

Breaking a mustang, particularly an older one, can take a fair bit of time and patience. One Idaho vet who has worked with scores of wild horses put it this way: "You stand to get your head kicked off dealing with those animals. Once I went out to castrate a stud and he just about tore up the corral by the time I got him sedated and tranquilized. He probably did two hundred dollars worth of damage before I laid a finger on him. Another time, I spent three hours standing around doing nothing until this mare calmed down. I gotta charge for my time, so when I work with a wild horse I charge by the hour. That bill adds up fast. The owner doesn't know whether to get madder'n hell at me or the BLM."

Which brings to mind a story I heard in Nevada from a cowboy who has spent half his life around horses. "They should *never* adopt a horse out after he's a four year old. No way in the world! All these people who are adopting a horse, ninety percent of them have never laid a hand on a horse in their goddamn life. They see this stuff on television about a guy going up and roping a horse and these people who adopt a horse think they can do that goddamn stuff. You know how many people have wound up in hospitals over horses they've adopted?

"For Chrissakes, there's horses running all over these hills with halters on them. They take a horse and they don't know what they're gettin' into. They put him in a chicken corral, the goddamn horse gets out and then before you know it he's up in the mountains. It's a joke, I'll tell you. It's a joke. No, if I had my way, I'd say a three year old. No older.

"Let me tell you about this girl from San Jose, California. We had a hundred horses out there at Palomino at the time. There were lots of nice little yearlings and she picked a strawberry roan mare that had to be twelve, fifteen years old. I went up to her and I said, 'Lady, why are you taking that mare? That mare'll kill you.' She said, 'Because she's a challenge.' I said, 'Well, she is a challenge all right.' So we run this mare down through this chute actually made to load cattle and we put a halter on her. That horse reached up and tried to bite the woman, just like that. Anyhow, she backed her trailer up into the chute and run that mare into it. This is a trailer that she rented in San Jose from a U-Haul outfit. A real nice, smart trailer. So we got that mare in there and she just jumped right up in the manger and just pushed it right down, right down into the saddle compartment. She kicked and kicked and all hell began flying loose. For Chrissakes, you should have seen it. There were pieces of lumber flying out of there this way and that way and everybody was kinda ducking around, trying to get the damn mare outa the trailer. She was skinned up badly, you know. She didn't belong in there. Finally, we got the damn mare outa there. So then I looked over and here's this woman sittin' down crying. I thought she got hurt. She was standing right by the trailer. I said, 'Are you hurt?' And she said, 'Yes, my *heart's* broken. I rented that trailer in San Jose and signed a contract that I'm responsible for all the damages on that trailer and all I've got is fifty dollars in the bank.' I stared at her and she finally said, 'The hell with that horse.' And she was gone just like that.

"Anyway, that's the kind of thing you run into. If they would take them horses over four years old and send them to the can . . . what the hell! We kill chickens, we kill turkeys, we kill sheep, we kill hogs, we kill cattle. What is the difference if a horse is killed humanely? And let me tell you something else. Fat horses from here are worth around fifty cents a pound right now. Fifty cents a pound! With that kind of money coming in, this horse program would pay for itself, and you as a taxpayer and me as a taxpayer wouldn't have that burden. If the people of the United States knew what this horse program is actually costing us, they'd think about it and get this goddamn horse law turned around. It's a damn shame."

Bill Stewart has similar feelings. "From my viewpoint, I don't think the BLM should process anything over seven years old. They shouldn't do it unless it's an outstanding horse, and you're then only talking about two percent. They shouldn't do it because these people who go in there and adopt a seven-, eight-, or nine-year old horse, they ain't gonna do anything with him anyway. Lot of them ol' studs'll kill you. The best thing is, you put these horses over seven in a big corral and when you get seventy-five or one hundred of them, call in the bidders. They're probably gonna go to the cannery. So what's the difference? The difference is, I'll tell you, is the horse program will operate and pay for itself. There's a lot of starving people who would eat wild horse. Look at the people starving in the old countries. Horses got some bone meal in them too."

While most who have dealt with wild horses seem to agree that adopting an older one is asking for trouble, a few dissent from this view. Andy Anderson, who works in the Carson City BLM office and is among the more astute horse experts within the bureau, claims that only one percent of the horses cannot be broken. He's clearly impressed with the plasticity of their behavior. He knows of several cases in which young girls and boys have had no trouble taming their animals. They discover that the first couple of days their mustang is fearful of a grain bucket, then before a week or two has passed it behaves as if the bucket full of oats were a natural extension of the muzzle. Anderson is not anxious to make a case against older horses. For every problem encountered he can cite an exception. For example, he knows a seventeen-year-old girl in the Carson City area who broke a ten-year-old stud in four hours. She was sitting on him the next day. The following year she was riding him in her high school parade. Before adopting her mustang she had no experience with horses.

Bob Moody takes an innovative approach to adopting out less desirable mustangs. Beginning in 1979, he ordered his Susanville cowboys to castrate all studs four years and older. Initially, Moody reasoned that since virtually all of the stallions would be used for riding, they would be gelded anyway. By doing it while branding and giving required vaccinations, the BLM could save the new owner between thirty-five and fifty dollars. He also found that gelding horses cut down on fighting. The results were so encouraging that by 1984 he was even castrating all his yearlings. Only when Moody knows beforehand that an older stud will be used for breeding does he allow the critter to keep his manhood.

He takes satisfaction in noting that his cowboys can brand and geld a horse in less than ten minutes, and without the use of an anesthetic. Veterinarians, he notes, only put a horse under to protect themselves from injury. They're not principally concerned with stress on the animal or the fact that now and again they'll lose one. Moody keeps records that back up his claim that he loses fewer studs than vets who insist on numbing the horse's nervous system. His losses, he claims, are less than one percent. Proud of his approach, Bob Moody says that the California Veterinary Medical Association once came to Susanville to check up on his method. "They were downright complimentary about what they saw."

To convince skeptics and curious visitors that his approach to castrations is quick and humane, Bob Moody had Allan Hoffmeister, the BLM's local public relations man, put the operation on videotape. All of it: the blood, the knives, the incisions, and plenty of closeups. While not something to be screened during the family dinner hour, the film shows that the premiering stallion hardly winces during the operation. After losing his baseball-sized testes in a couple of decisive slices and then receiving a tetanus shot, he's up and running about in a matter of minutes. Galloping, whinning, snorting, and kicking high.

Bob Moody explains that the Susanville operation has been unique in other ways. Elsewhere, the BLM uses one crew for the roundups and another one for work at the corrals once the horses are brought in. "Other adoption centers, they've got people working part-time or sitting around wondering what to do next. That's inefficient, not good for the morale." His cowboys work on a well-defined cycle. First they participate in roundups for several days. Then for several more, they sort, brand, vaccinate, halter, and geld. And then, in what Moody describes as the third phase of the cycle, one and all pitch in to help mother, brother, dad, and sis pick out the right horse. Moody has also used his cowboys to round up burros. Between 1980 and 1984, Moody and crew gathered almost ten thousand of them, most in Death Valley National Monument and California's China Lakes Naval Weapons Center. About half of the burros have been adopted out through humane groups, the rest by the BLM or its contractors.

Some years ago, to the consternation of other BLMers associated with the wild horse program, Moody started his own rodeo. It's still the only one of its kind in the nation. He built public bleachers at the government corrals in Litchfield, about twenty miles east of Susanville. Anyone is welcome to tour the facility's forty acres and watch his cowboys head and heel a large dun stud, tag and halter a spry mare, or brand and vaccinate a colt that seems lost and frightened half to death. The rodeo is for show and for advertisement, and Moody admits that a certain amount of theatrics is encouraged. "You want to know why I do it?" he asks rhetorically. "Because these are hard economic times. You're gonna put shoes on the baby before you're gonna put a horse in the backyard, aren't you? You betcha."

Whatever the real payoff from Moody's rodeo, it stands in stark contrast to the all-business, no showbiz approach in Nevada and Wyoming. There one finds no bleachers and no signs welcoming the public to watch the processing of mustangs. About the biggest treat you'll get in Rock Springs, Wyoming, is the sight of a couple of wranglers chasing a half-dozen stallions from one corral into another. Much of the rest a spectator probably doesn't want to see—not unless he's got a sadistic streak in him. Once chuted, the outsized Wyoming wild ones bang and kick and bruise and pile up on one another. Not infrequently, a mare or a yearling gets clobbered or squashed between a couple of rambunctious half-ton studs. The victim might find itself racked between steel chute bars five feet off the ground, helplessly splayed like meat on a butcher's block. The predicament doesn't improve much when the horse gets locked in and surrounded by jittery pairs of human hands. Gloved claws that lunge and probe, shave and paint, cut and stick. Blood is drawn, worming paste injected, the mane shaved, and a complicated freeze brand applied. Minutes later a cane is jabbed into the mouth in a dubious ageing exercise. Then, an attempt is made to halter the animal. It kicks and bucks for all it's worth, and it contorts its face: it seems to be saying that having to wear a vulgar leather rig is the ultimate indignity.

I asked Bob Moody about the often-voiced accusation that he was merely running a show, and that his cowboys only rope the horses because it gives them pleasure. I asked if, on balance, his methods were more humane than those I'd seen in Wyoming. He laughed to himself, as though he had heard these questions so many times that he had lost interest in giving answers. He scratched his vacant pate, ran a hand through the black scraggly hair around his ears, and for a brief moment I saw him as a onetime Hashbury hippie. Finally, with a broad smile, he exclaimed, "Hey, you gotta do what you can, what works! Besides, headin' and healin' ain't no harder on the horses than shovin' them into chutes. Maybe hard but not harder. It's easier on them."

Is it easier on them? Moody's headin' and heelin' demonstration has met with criticism from several wild horse BLMers, and in July 1984, a BLM employee named Sharon Saare was asked to compare Moody's method with the squeeze-chute approach used in Nevada. Ms. Saare tested twenty-four horses, twelve in Susanville and twelve at Nevada's Palomino Valley. Results were mixed. For some horses, the squeeze chute was more stressful; they had a higher pulse rate than those that were roped and dropped and branded on the spot. Other mustangs remained calmer in the chute.

Moody and those under him go out of their way to get people to take burros and mustangs. In the weeks prior to a gathering, BLMers contact prospective adopters and set a firm date for them to pick up their animal. "If we don't have a definite adoption source for 'em, we don't gather 'em."

Fortunately, Bob Moody has been able to call on the talents of Allan Hoffmeister. A lean, affable redhead, Hoffmeister spends a disproportionate share of his public-relations time advertising the adoption program. He has tried just about all the angles that might occur to an up-and-coming ad man. He does media blitzes and prepares brochures, black-and-white photos, slides, and videotapes. He films the cowboys on roundups, playing rodeo, feeding and caring for sick horses, helping families become foster parents. Film cans in hand, Hoffmeister then hustles television stations in Sacramento and the Bay area to show his videotapes on the local news. He identifies volunteers to take telephone calls, sends out qualification applications, and answers questions about when the horses can be picked out, picked up, and taken home. When he's on the road in northern California, Hoffmeister also searches for livestock yards and county fairgrounds where the BLM can set up temporary facilities for a couple of days. Gratis, if you don't mind—it's a worthy cause.

Despite the genuine efforts of many Susanville BLMers, they have all discovered that people think twice before dishing out the equivalent of a good night on the town for a captivating part of our national heritage. According to Hoffmeister, almost half of those who see the television tapes and make a call and fill out an eligibility form fail to pick up their animal. "And it's darn frustrating when you've done a lot of work and asked others to donate their time," he says. "What we're going to do to get rid of some of these animals I really don't know."

Moody has had no desire to tell the public that if the BLM continues having problems getting rid of horses they will have to be killed.

"Why not?" I asked.

"Because we can *do* it!" he exclaimed, his salt and pepper mustache kicking up a storm around his mouth. This kind of positive go-get-'em attitude is Bob Moody's halo, his philosophy about everything he gets into. And yet he appreciates that there is a darker side to the Adopt-A-Horse Program. He has been with the government long enough to know that administrations are fickle and the fate of his best intentions is determined a continent away. "You want to know what my problem is? My limiting factor is dollars. The country won't fund the government to do much of anything, especially since the Reaganites got in. What's so devastating about horses and burros is that you can lose everything you have gained by your inactivity because they multiply so fast. So if you're inactive for a year or two because of cutbacks, then you've lost everything you've gained in the past. This job, if you want to know the truth, is a continual battle to try to get the people and the resources and the dollars to get the chore done."

Just before leaving Susanville, the electric Moody said, "Did I ever tell you about the vet from the University of Nevada who was up here showing some of our crew how to age a horse? He put a finger in this stud's mouth and it came out minus a digit. So just like nothing happened he picks it up and puts it in a handkerchief and decides it's time to go home. Just like that!"

Mike Zeis also does what he can to find good homes for wild horses and burros. Unlike Bob Moody, however, Mike works for himself. Or rather he works on the family farm, and as an adoption contractor for the BLM.

When Mike was about to graduate from the University of Nebraska, he saw a newspaper ad announcing that the BLM wanted to open an adoption center for wild horses and burros within fifty miles of Omaha. The government, the ad said, was accepting bids and proposals from anyone who thought he had entrepreneurial skills and enough land for corrals. Mike had first heard about the adoption program several years before when he wanted a burro for breaking calves. He decided not to get one from the BLM, however, because he'd have to travel to Arizona to pick it up. When he accidentally happened on the BLM ad, his interest in the adoption program was revived. Since the family farm is near Valley, twenty miles west of Omaha, he reasoned that if he could win the bid he could use his book knowledge of advertising to good advantage. He could also bring in much-needed income to keep the family farm going. He wrote to the BLM for information, and shortly got more than he bargained for. "I laughed for a day and then some at the stack of paper and rules and regulations they sent," he now says. The disorganized mountain of words did not deter him, however. He got down to business and spent much of his last two months at the university preparing a proposal and a required environmental impact statement.

When Mike and his father met with BLM interviewers—they were in competition with twenty-seven others—they were asked why they wanted the center. "We want to be honest about it," Mike told them. "We're in this to make a buck." To which his fragile-looking father added, "We can't donate any time to anybody the shape we're in. We're farmers; you know what that means." By the time they were finished Mike had apparently impressed government interviewers. In August of 1981, he learned that he had submitted the winning bid.

Among other things, Mike learned that he had sold the government on his advertising plans. "I took the position with the BLM that this thing wouldn't work unless it was actively promoted," he says. "The BLM does an extremely poor job of public relations. I spend probably eighty percent of my time doing promotional work. Press interviews, newspaper articles, talk shows, local radio and television. I have films I show to grade schools and high schools. Boy Scouts and Girl Scouts and 4-H groups come out to see the animals. I send out releases all the time to small and large towns in Nebraska. Without this kind of work I wouldn't have moved half as many animals as I have."

Mike Zeis is one of a half dozen or so individuals around the country who have contracted with the BLM to adopt out horses and burros. These adoption centers—in Pennsylvania, Tennessee, Oklahoma, Texas, and the one in Nebraska—supplement those run directly by the BLM in California, Oregon, Nevada, and Wyoming. Mike's center may be the most successful of the privately run operations. In the first four months after opening, he found homes for five hundred horses and three hundred burros, and then moved another two hundred animals during the normally slow winter months.

Mike humorously complains that adopting out wild horses and burros is a much more time-consuming job than he initially envisioned. With voluminous paperwork, feeding and caring for the animals, catering to questions and problems from adopters, and the work involved in advertising, he rarely puts in less than forty hours a week. He has discovered that weekend travelers love to stop just to see what a mustang or burro really looks like, and they have a penchant for asking more questions than he's got answers for. Why does he continue with the adoption center when he could obviously make more money by trading on his college degree in the commercial world? "Because the experience is worth more than five years I could be spending with an Omaha advertising firm," he says with considerable assurance.

Taking a contract with the government has also allowed him to learn about its annoying bureaucracy. He says that his phone bill occasionally runs more than $400 a month, mainly for calls to Rock Springs to find out when and if his next shipment of horses will arrive, and to the Denver Federal Service Center which holds the nation's statistical records on wild horse and burro adoptions.

Mike is certain that every unflattering charge ever leveled against government bureaucracies is true. He's got a garbage can full of examples to prove his point. For example, while it shouldn't take more than three or four weeks for adoption applications to clear the Denver office, after which time people can pick up their horses, some of his customers have had to wait two or three years to get their animal. "Things always get jammed up in Denver," he says. He likes to cite the time the Denver Service Center ran out of computer print-out paper and it took a full six weeks before anyone found a supplier. Without the proper paper no one could be notified that his application had been approved! Another difficulty was, if nothing else, unusual. Mike sent in a batch of adoption applications on behalf of eager adopters, feeling that this would speed up the process. When they weren't returned after several weeks, he called to find out what went wrong. The applications, he was told, had been mysteriously lost, or rather, they had been sent to a prison to be key-punched. No one knows exactly what happened at that point. One rumor had it that the prisoners made paper airplanes out of them.

At one point in 1982, the national total of those who said they wanted mustangs dropped from eighteen thousand to five thousand. It was claimed by some in the BLM that this resulted from the eight-fold increase in the adoption fee, from twenty-five to two hundred dollars. But did it? Those processing applications in Denver had sent all applicants a letter informing them of the fee change. Those who still wanted a horse had thirty days to sign and return a form affirming their interest. No one seemed to realize until too late that people don't like to reply to much of anything. No one saw that it would have been a whole lot better to have applicants not reply if they still wanted a horse. As a result, everyone who didn't respond within thirty days was dropped from the "I-want-a-mustang" list.

Notwithstanding these irritations, Mike Zeis sees his Valley adoption center as a major success. "Horses are a distribution problem, and distribution is what advertising is all about. If you advertise properly, you're going to move your product." True enough, but it's also true that the Valley adoption center has a couple of things going for it. For one, it sits in the middle of a pretty good market. It is the only adoption facility between Rock Springs, Wyoming, and Cross Plains, Tennessee, and though many of Zeis' customers are Nebraskans, a good many drive in with empty trailers from out-of-state: Iowa, Missouri, South Dakota, Minnesota, Kansas, Wisconsin.

Most of Zeis's wild horses are adopted for use as saddle horses. The biggest ones he gets are twelve hundred to thirteen hundred pounds, and they go first—for driving teams. He says that with every load of horses he gets from Wyoming, he can usually find a team that is well-matched for size, color, and age. "And some we have adopted have been so well matched they've appeared in parades and at state fairs." Some of the horses are taken for use as pack animals, others for hunting trips in the Rockies. He even had the Nebraska Game and Parks Commission take eighteen of his mustangs for use in tourist trail rides and in the state's several parks. Now and then an older mare will be adopted by someone who wants to raise colts.

The demand for burros is very strong, particularly from farmers in Iowa and Missouri. When he receives a load of seventy or eighty of them, which he gets from California and Arizona, they are all claimed within three or four days. "Unlike the horses, they walk right up to people, and first thing you know they're gone." Many people want burros for breaking calves or tilling gardens, some for riding, a few just want to caress them. The specifics on who takes the dusty beasts of *Death Valley Days* fame are varied and suggest the resourcefulness of adopters. One rancher in western Nebraska adopted two burros because he has a gold mine in Colorado. To keep his claim legitimate, he has to take out two hundred dollars of ore a year. The mine is only accessible on foot, so he uses his burros to pack in supplies and carry out hope-filled rocks. A farmer–rancher from Arkansas adopted several burros to raise with his sheep. This yeoman from the Land of Opportunity had heard that burros imprint on sheep, and then forever after protect them from coyotes and dogs.

Even Mike couldn't resist the temptation to adopt the endearing burros. In Susanville, where Mike is a favored customer because he is willing to take older burros, they were happy to let him know in early 1983 that they had a rare pair of paint jacks. The jacks, which now pull a small hay wagon on the Zeis farm, are Mike's pride and joy.

The wild horses that pass through the Valley adoption center are, by objective standards, among the biggest and most handsome to be found anywhere. They all come from southwestern Wyoming. Zeis says of his horses, "They're not like those Nevada beasts. Everyone I talk to agrees those Nevada horses are smaller, skinnier, uglier, and harder to get rid of." Zeis not only draws exclusively on the best-bred and best-fed mustangs in the nation, but he has somehow managed to get the BLM in Rock Springs to send him the best of the best. He says that virtually all the horses he receives are between one and four years old, and the majority between one and two. "I tell the people in Rock Springs I have to have young ones to move them. This is big 4-H country. People in Nebraska want young females and they want a quality horse."

In the mid-1970s, people in and around Reno, Carson City, and Gardnerville adopted more mustangs on a per capita basis than any comparable group of cityites and suburbanites in the country. In one four-year period, border-hugging Nevadans adopted almost six hundred mustangs. Transients, job transferees, and the disaffected were cashing in on a dream: owning two acres and a horse within an hour or so of some of the best skiing in the world.

No doubt adopters had good intentions and an abundance of loving thoughts. But for many of those who come and go in the gaming, hotel, and service businesses of Nevada, and for some longterm residents, dreams are fragile. Seemingly tender concerns for animals are rouge-deep, particularly when mustangs can be had so cheaply. As it turned out, adopters in no area in the nation required so much educational work by the BLM, who had to tell novices how to bridle or ride or corral or feed or just plain care for a horse. Worse, when adopters received pink slips at work or decided to move on to new employment, many just opened the gate and let their horses go. Or, if their conscience got the best of them on the way out of town, they called the BLM to say that there was an unmanageable animal in a corral on Carefree Lane and it would be a good idea if someone came quickly to pick it up. By one count, there have been more than a hundred of these open-corral or come-and-get-'em cases in recent years along the Reno–Gardnerville axis. In Lemmon Valley alone, a sprawling, countrified suburban tract between Reno and Palomino Valley, a good dozen mustangs now roam within shouting distance of the newest homes. There was none there prior to the time horses became available for adoption.

One of the areas that experienced rapid suburban development in the 1970s was the western slope of the Pinenut Mountains. Reaching elevations above 8,000 feet in its eastern reaches, its rolling foothills south and east of the Carson River offer attractive vistas. And, in addition, bands of cinematic mustangs. Eight hundred to nine hundred wild horses still roam the rugged Pinenuts, as they have since the nineteenth century. The horses are a common sight in winter, when food is scarce at higher elevations. In subdivisions near Dayton and Carson City, just about everybody was initially impressed with the horses. After all, what could be more satisfying than to sit down for dinner and see, not two hundred yards away, a gathering of five or six stately looking mustangs on an arching, pinyon-mantled hill.

Soon, however, people became dissatisfied with the distant image of a photogenic band of mustangs. Many began throwing out bales of hay to bring them closer, thinking that it wouldn't be long before they could stroke an untamed muzzle or hand-feed a darling colt. Some foothill residents even left fragrant hay exposed in their yards. Sure enough, the mustangs came to feed and they began to lose their wildness. Now people could get within twenty-five to thirty feet to take pictures, to chatter affectionately at them, to challenge and chase.

Ferreting noses and eyes and primitive instincts moved into overdrive. The mustangs trampled through gardens, they ate fruit trees, and they mowed lawns better than the most fastidious home gardeners cared to imagine. The four-footed visitors had a field day chomping on carefully tended flowers of every description. No one knows this better than a group of Colorado biologists who did fecal analyses on one hundred fifty of the Pinenut horses. They were startled, then positive they had been sent the wrong samples. The wild horse droppings were full of more than a dozen non-native plants.

Bill Stewart, the same Bill Stewart who was once in charge of Palomino Valley, owns an attractive home, a roping school, a couple of cows, and half a dozen horses on several acres overlooking the Carson River. Over the years, he has taken three mustangs home for family pleasure. When we talked at his home, he recalled how one of his neighbors put in a distress call to him about an ill-mannered mustang. It seems the woman had a terraced garden full of tomatoes, radishes, peas, squash, and a whole lot more. "So I went over 'cause that horse was right in the middle of her garden eating that nice wet leafy lettuce, just like he owned the place. I did what I had to, let me tell you. I roped him and got him out of there, just like that. But I knew he would be back in a day or two."

Sitting on Stewart's front porch drinking wine one evening we saw, not fifty yards away in ungainly tall sage, a small gathering of mustangs: a plump white stallion, a bay mare, a dark yearling, and a colt hugging his mother's flank. Before I had identified the last of them, Stewart shot forward in his chair and cried, "I could jump out of here and rope one of them and put him in a corral before you'd even know it. Why, in a year you wouldn't even know he'd been a mustang!"

I asked him if he had much of a problem with the horses coming into his yard.

"Let me tell you something," he said. "I'll put a sack of hay right where that pickup is and I'll guarantee you within a week you come out here one, two 'clock in the morning, there'll be wild horses eatin' it. Now you're probably wondering why the horses don't go after the hay right away. I'll tell you. They're used to natural grasses. That hay's just loose stuff and they'll lay in it but they won't eat it right away. Those wild horses they got to be careful. If they don't have enough moisture in them to digest the hay it will ball up on 'em and kill 'em. But like I was saying, just come out here and turn on that yard light a week after you put some hay out, there'll be wild horses going after it. Ain't that right, babe?" he said to his wife, Jan, standing near the door. He answered for her: "That's right."

Nothing is quite so inviting to a stud as a lonely mare. To the strong and mighty, there is no such thing as too many wives. Little, of course, so angers the two-acre homeowners as the sight of their prize thoroughbred or quarter horse being pursued by alien, inferior genes. One subdivision resident near the Carson River, who lives near the Stewarts, went into a rage when he saw a wild stallion mounting his registered Arabian mare. In another case, a golden retriever thought he would protect his owner's assets. The challenged stud stomped him to death.

Indignant that the horses were boorish and ungrateful, many families put up electric wires or fences, sometimes no more than picket fences. But mustangs don't care about such human artifacts, and it takes only a minute or so for them to crash through prettified garden lattice. A few people apparently thought they could solve their problems by scaring the horses with a good chase on a motorcycle. No luck. A few more decided to define the law in their own terms: they shot horses—thirty-five to forty by 1984. In one better-forgotten instance, the killer slaughtered a mustang and then threw pieces of the carcass here and there, as if part of a dark Stygian ritual. Gut-shooting became the tortuous form of vigilante justice. A horse is shot in the stomach with a low-caliber gun; it then painfully wanders off to a slow death, often miles from the scene of the crime.

People wanted something done about the "problem horses." There were demands that they be removed, equally vociferous counterdemands that they be left alone. Citizen groups were formed; petitions were signed and sent to the BLM by both factions. The BLM began getting calls that the horses were invading one of the golf courses, that some were taking evening strolls along the highway. Then, as if all this wasn't enough to drive local BLMers crazy, Washoe Indians complained that mustangs were breaking down fences and bothering their saddle horses and making it difficult to run cows on their scattered one hundred sixty-acre parcels in the Pinenuts. The Indians had themselves to blame for much of the trouble, however. When they acquired tractors they released many of their horses, figuring that they'd recapture them if they didn't like their expensive acquisitions. But they did, and they forgot about their once-domestic mounts. Before long, the horses had turned wild and were soon multiplying with abandon.

With all these complaints, the BLM had to do something, and that something was to begin removing horses from the Pinenuts at the rate of about fifty a year—more than three hundred fifty by 1984, not counting the two hundred horses that the BLM removed on behalf of the Washoe Indians. But removing half a hundred mustangs a year from the expanding suburbs may not be enough. The BLM continues to receive numerous complaints from the highway patrol, and from all kinds of people, including those who had previously signed petitions to leave the horses alone. It was becoming abundantly clear that many had said to themselves: "To hell with the romanticism of mustangs. Give me my private property and my hard-won domesticity."

Congress declared that the West's mustangs are a "national heritage," and its 1971 law implied that they were to be enjoyed in the way some of us enjoy our national parks and monuments. Yet, relatively few people make any effort to see the mustangs. Are most content to know that they exist? Are many people simply ignorant of their existence? Or don't they know of the horses' exact whereabouts?

The BLM has done very little to make the horses accessible for public appreciation, or the public aware of how to get to them. Recently, the Nevada BLM has said that horses left in Stone Cabin Valley on State Highway 6 between Tonopah and Warm Springs are there for public viewing. Their rationale, I suppose, is that this is one of the few places in the state where one can almost be guaranteed of seeing mustangs almost any time of the day. And, to boot, along a well-paved highway, where you don't even have to slow down or use binoculars. The only problem, beside the fact that Stone Cabin Valley is about as far off the beaten track as you can get, is that the horses there look like a so-so bunch of rather tame ranch stock. Or, if you use binoculars, what you'll see are anything but Zane Grey's statuesque stallions at a command post on a high butte.

Once, I asked Andy Anderson in the Carson City BLM office what the bureau was doing to help people see mustangs. He reminded me that within a few miles of where we were talking, tourists could see all kinds of wild horses if they so desired. He said that recently a few people had come by and wanted to see them. He told them where to go, they saw horses, and then they returned to express their disappointment. "Is *that* what they look like?"

What, it might be asked, is on people's minds when they think of wild horses? Mark Rey, in a master's thesis done at the University of Michigan in 1975, is apparently one of the few to have asked the question. In a survey of more than two hundred people, he found that thirty percent thought that merely seeing mustangs running free justified their place in nature. Another thirty percent thought them beneficial because "they are associated with America's heritage," or "they have educational value," or "because I know that some are still left." When these same people were asked what feelings they associate with mustangs, forty percent said "freedom," another fifteen percent said, "spirit, beauty, and majesty." In the minds of this small and unsystematic sample, only bighorn sheep and elk were thought to be "more exciting to see." Wild horses were judged to be aesthetically no different than golden eagles, and significantly more interesting to contemplate than mule deer, red fox, great blue herons, and coyotes.

According to Rey's survey, a fair number of people have said that they are willing to hike, or drive over bad roads, to see wild horses. They clearly don't want their vista marred with human artifacts: homes, fences, power lines, well-paved roads. It seems that the greatest number might be happiest with auto overlooks, perhaps a Grand Canyon-like panorama of scores of many-colored mustangs playing, fighting, eating, feeding young, spontaneously running free and mingling with nature's natives. In all, stirring one's sense of the West, of unbuttoned independence, of creation provident and divine.

From 1971 to 1975 Dawn Lappin, then Wild Horse Annie's chief lieutenant, was (as she tells it) the key figure in the West's wild horse adoption program. "In 1972, the BLM said an adoption program could never work, and *I* made it work. In 1974 they asked me to place four hundred horses and said it couldn't be done, and *I* did it. In 1976 they said we had saturated the market. Just *look* at what happened. See how many have been adopted out since 1976."

Dawn Lappin's experience taught her that, on the whole, people have taken wild horses because they've been cheap, and because they were called mustangs. Bad conformation, a misshapen head, size, questionable health—all these issues have been secondary.

When, however, Robert Burford, national BLM director, raised the adoption fee from $25 to $200 at the beginning of 1982, secondary concerns became primary. As Dawn says, "People didn't want to pay

$200 for a horse for which there was no guarantee that it was even going to live the first two days. You know, I worked real hard for ten years to adopt out horses in those early years and bring about what I thought was good management of the horses, and then when the BLM raised that fee, I saw ten years worth of work crumble down to the floor."

Even before the new adoption fee became law, all kinds of people, including a great many quiet ones in the BLM, opined that Burford's effort to find a way to make the horse program pay for itself would cause irreparable damage. Many simply couldn't see the logic. One Oregon BLMer said, "That $200 fee, it's dead wrong. You wouldn't pay $5,000 for a car that's only worth $2,000, would you? Now with this new ridiculously high fee we're going to have excess numbers, no way around that. What are we going to do with them?"

Organizations such as the Humane Society of the United States and the American Horse Protection Association vehemently protested the new price. Before the fee was even tested in the marketplace, the AHPA had sued James Watt, Secretary of the Interior, to prevent the destruction of horses that couldn't be adopted. The lawsuit contended that the BLM's actions were "arbitrary, capricious, and violate the Wild Free-Roaming Horse and Burro Act." The suit also claimed that the bureau was creating a need to slaughter mustangs. The Animal Protection Institute in Sacramento, California, claimed that the new fee would dry up the market and somehow lead to the quick extinction of the West's wild horses. Within the first six months after the new price went into effect, the BLM in Washington, D.C., received over eleven thousand letters of protest. The letters came primarily from elementary and junior high school students. Letters written by girls outnumbered those from boys by seven to one.

The eight-fold increase in the adoption fee had clear-cut effects. In Nevada in the fiscal year 1981–82, forty-one hundred wild horses were rounded up in the state and either adopted out directly from Palomino Valley or sent to other adoption centers. But during the following year the comparable roundup figure was twenty-five hundred. Four hundred of these horses spent an inordinately long time in government corrals before they were finally put up for adoption.

Consideration was given to shooting unadopted horses after they had been held for forty-five days, as permitted by law. But deliberations went nowhere. The Nevada BLM was acutely aware that, among others, Dawn Lappin and the ten thousand members of Wild Horse Organized Assistance which she controls could have bureau chiefs walking on hot coals. Lappin was so unhappy with the fee increase that a month after it took effect she sent the Nevada Director of the BLM, Ed Spang, a check to care for eighty horses held in Palomino Valley. The amount was the equivalent of what a rancher pays to keep the same number of cows on public land for one month. When Spang refused her offer, Lappin enlarged the check bigger than a horse's back and delivered it to him personally. Spang and others in the BLM correctly perceived that the political fallout from killing the horses would be more than they wanted to cope with.

Why was the fee raised to two hundred dollars? According to John Boyles, Washington-based chief of the entire wild horse program, David Stockman, Director of the Office of Management and Budget, told Robert Burford in late 1980 that the wild horse program would, like other federal programs, have to be self-supporting. Budget analysts claimed that it cost $535 a head to round up and care for the horses until they were adopted. This, it was declared, was what adopters should pay for a mustang. Burford balked at the high price, sensing that it would bring the adoption program to a halt. They eventually compromised, and the fee was set at two hundred dollars.

Dawn Lappin didn't see David Stockman as the culprit. She blamed the fee increase on Robert Burford and Interior Secretary James Watt, believing that "Watt didn't have the foggiest idea what the wild horse adoption program was all about." She reasoned that Burford and Watt, buddies from Colorado, had hatched the higher adoption fee in order to eventually dispose of the entire adoption program. "If necessary, Burford would raise the adoption fee to $500 to get rid of the program. And then the BLM can go into the horse production business."

"You don't really believe that?" I asked.

"Oh yes, you bet that's what they wanted to do." As evidence Dawn will cite an unnamed BLM district manager who told her that high on his list of priorities was a desire to "upgrade" the mustangs. "Because then the BLM could get more money for them." She says she told the district manager, "I don't care what those horses look like. They can look like a bale of hay for all I care. All the kids care about who adopt a horse is that it has four legs and can be called a horse."

The Washington bosses, according to a couple of small fish in the BLM, had other reasons for the manyfold increase in the adoption fee. One was that a price tag high enough to be felt would winnow out those enthusiasts who don't realize that adopting a large, untamed animal is a serious commitment. If indeed this was central to their reasoning, then the thought was a good one. Those who deal in the sale of animals find that people who pay little or nothing for them provide poor care.

At the time of the fee increase, a widely held view among many horse protectionists and some in the BLM was that the $200 fee was really a ruse, a measure needed to ensure the BLM's right to sell the horses commercially. The government, so the argument went, knew that a much higher price for a horse with generally poor conformation would drastically reduce demand. Then, to no one's surprise, mustangs would begin accumulating in government holding facilities. With a daily food and boarding bill of a little over two dollars per horse, it wouldn't be long before the cost of the adoption program looked grossly out of line with conservative government thinking. At this point, the Reaganites, despite coast-to-coast protests that wild horse activists could drum up on short notice, would get what they wanted: new legislation to dispose of "unadoptables."

And so it was. Once the adoption fee was changed, Burford tried to garner support to amend the federal law so that unadoptable horses could be sold commercially. Money received from the sale of horses could be returned to those states charged with managing the animals. Backing Burford were two U.S. Congressmen, Representative Don Young from Alaska and Senator James McClure from Idaho. The Young and McClure amendments informed Congress that noble though its intentions were in 1971 when it enacted a law declaring the wild horse a "national heritage, a living symbol of the historic and pioneer spirit of the West," the West's wild horse herds were thriving and by no means in danger of disappearing. It was time, they said, that the federal government give true meaning to a multiple-use concept of public rangelands as recognized in the Federal Land Policy Act of 1976 and the Public Rangelands Act of 1978. Multiple-use, under these laws, means that cattle, sheep, and wildlife have as much right to public lands as do wild horses and burros. The Young and McClure message was clear: reduce the mustang population and thereby give western livestock owners what they are clamoring for—a population of wild horses no larger than the 1971 purported figure of seventeen thousand.

Dawn Lappin says that she could have lived with an increase from twenty-five dollars to one hundred dollars. "At least it would have kept out all those who only take a wild horse because it's the 'now' fad. At $100, I could have adopted out eighty percent of them. And the other twenty percent, they're not adoptable anyway. They're too old or sick or something's wrong with them." These leftovers, she says, could be sold commercially.

Dawn would have gone for other possible solutions: for example, set a price for each horse based on its quality, or base price on the weekly market rate for domestic horses. Here her thinking was similar to that of two Reno Sierra Club members active in the wild horse program, Rose Strickland and Tina Nappe.

Instead of a single price of $200 for all horses, Rose and Tina have advocated that horses be priced according to quality and an index that reflected prevailing market prices for comparable horses. Those that are big and beautiful and have good conformation could be priced at two hundred dollars or more; the price for others could be scaled down accordingly, some undoubtedly selling for well under a hundred dollars. Actually, Rose and Tina had also thought that a two-stage auction plan might be feasible. In stage one, the general public would have the first opportunity to bid on unadoptable horses; presumably, many would be taken as pets. Those remaining would be sold in a general auction. All money received from these sales would go back into the wild horse program.

By 1983, adoptions were moving slowly and Robert Burford was getting heat from all quarters. Finally, in March of 1983, he ordered the adoption fee reduced to one hundred twenty-five dollars. It was a victory of sorts, as the number of people willing to take a mustang again picked up. On the other hand, there were unmistakable signs that the adoption program was now well beyond the honeymoon stage. The rate at which horses had to be gathered just to keep up with annual increases in the population was not slowing. The BLM was having to make greater use of satellite centers—new locations that would be used for only one or two weekends a year—in order to tap a fresh market.

By mid-1984, another problem was apparent: old horses, and even younger ones, were accumulating in government corrals. In Rock Springs, where adopters invariably find the most handsome and robust mustangs in the country, there were several hundred on hand despite efforts to give them away in large lots to humane organizations and Indian tribes. Susanville's adoption center had more than two hundred horses that were seven years or older. Fifty of them had been on hand more than three years. The problem with old horses was worse in Nevada, where there were almost four hundred fifty "seniors." This number included three hundred sixty studs and fifty dry mares. About the best solution those at Palomino Valley had been able to come up with was to give a free foal to those willing to take an old mare.

CHAPTER 4

RANCHER WOES

He was sitting at the far end of the counter, finishing his breakfast. I sat down next to him and had a cup of coffee. We chatted briefly, then I asked, "Where's the best place to see wild horses?"

"Stone Cabin Valley, east of Tonopah. No place like it anywhere."

"Can you see many from the highway?"

He laughed jauntily. "You bet. Every last one of them. Hell, they got so many down there you could be a millionaire if you rounded them all up. Every year they're doubling. There's now two hundred fifty thousand head of wild horses in Nevada."

"Is that right?" I replied, knowing full well that his number was the grossest of exaggerations. At most there were forty thousand to fifty thousand in the entire state.

"You ever hear of Railroad Valley?" he asked.

I shook my head.

"Right near Stone Cabin. Railroad had better than twenty-five thousand head of horses all by itself in the 1950s. I remember you could drive across the road and see three hundred, five hundred in a bunch. Everywhere you looked you could see dust and clouds of dirt. Every place you'd go, not just one place, there would be a bunch of horses. You seen nothing unless you seen Railroad Valley in the 1950s.

"Why, there was another valley—I don't right off remember its name—it had fourteen thousand head of horses in that one valley and eleven thousand in another one nearby. Man, I talked to this BLM guy and told him if they don't do something real quick, it'll be another three years and there won't be a single blade of grass left for a cow in the state. Why, I'll tell you something else. I had a friend he went out and counted eight thousand head near his ranch and he rounded them up and rounded them up and out of that he thought he'd get eight saddle horses. He had to get rid of four of them. They was that bad."

He tugged at his visored cap and then went on. "Each one of them wild horses eats enough to feed four cows. Four cows, you betcha. Them horses are so goddamn hungry they're out there at midnight eating that cow feed. I'm from down around Goldfield and . . . have you seen all those hungry horses down by Goldfield? Down there in the Stonewall Mountains? Those horses eat a bale of hay a day. It takes a cow three or four days to eat one. Four goddamn cows to one horse, and there's two hundred fifty thousand of them in the state.

"What the hell this governor and the BLM is going to do about it I sure don't know. You know, in this state if you want to get something done, you gotta form a board and put three BLMers on it and six ranchers and then you'll get something done. And that's a maybe."

"I take it you're a rancher?" I said.

"No, I just goddamn know about these horses."

When the stranger left, I tried to decide if anything he had said contained an element of truth. About all I was able to conclude was that one of Nevada's state senators, Clifford Young, had been right on target when he recognized that wild horses have a genuinely glandular effect on people. In 1974, Young proposed to his legislative colleagues that mustangs be designated the state animal because of their remarkable kicking and fighting ability, and because they exemplify "the spirit of independence and adventure that is characteristic of the State of Nevada." And then Young added that the horses have such phenomenal breeding habits and such an unparalleled ability to denude the range that they could profitably be used as a military weapon in Vietnam!

With less than two hours of sunlight left and the sky a dappled grey, I left Tonopah and drove east on Highway 6. Forty miles down the road, I turned north into Stone Cabin Valley, which opens into an irregularly sloping flat that stretches like a gargantuan stage northward between the imposing Flat Creek Range on the east and the more gentle and distant Monitor Range to the west. The dark sky and the sharp angles of Rawhide Mountain in the Flat Creek Range made me think of a Stephen King setting. In the middle distance, I could see a luminous clump of green, which I guessed was the oasis where I'd find the Clifford family, present tenants of the one hundred twelve year-old Stone Cabin Ranch.

When I saw no sign of horses, I pulled over to the side of the road, got out, and began to carefully inspect the ground. Everywhere I looked the earth was dry, beaten, bereft of better grasses. I strained to find small bunches of needle-and-thread grass, sand grass, wheat-grass, others that I had come to know. I stared at sweet smelling sagebrush and noticed that this woody muscle of the desert looked hammered, chewed, cowering. Somebody or something was taking unfair advantage. Before long, it began to register that unusual amounts of halogeton and other weeds were everywhere. A large, noxious weed with blue-green, sausage-shaped leaves, halogeton is commonly found in disturbed western rangelands. Halogeton made its first appearance in Nevada in 1934, coming originally from Siberia or China. It prospers in areas with little rainfall and alkaline soils, especially where there has been overgrazing. Although full of miner-als and more crude protein than alfalfa, it is poisonous to cattle and sheep. Wildlife and domestic stock eat halogeton when nothing else is available.

I got back in my truck and took the first dirt road that headed off in the direction of the Monitor Range. I hadn't gone more than a mile along a trail overgrown with rabbitbrush when I saw a large sorrel stallion in a clearing below low hills off to my left. His ladies and one colt stood nearby. With the exception of the colt, who was nuzzling up to his mother, all of them were intently watching me. Cautiously, slowly, I got out of the truck and tried to approach them on foot. But when I'd gotten to within a hundred yards or so, the stallion snorted and whinnied. I paused; he stared and raised his head. I waited a few long moments, then slowly tried to close the distance between us. He pawed the dirt, lowered his head, then swung around and flattened his ears against his neck and moved several steps closer to his family. He was sending an unmistakable signal that I was trouble, that it was time to leave and finish the evening meal elsewhere. I didn't move, but now it no longer mattered. A bay mare hurried around to the far side of the band, slowed to a walk, and without looking back led the way toward higher ground. The colt and the others had gotten the mes-sage. The stallion momentarily remained motionless, his eyes fixed on me. Then, without further ado, his harem safely on its way, he turned and followed in a slow gallop. Soon, he closed the distance and brought up the rear of a line as straight as an arrow. The seven mustangs topped a rise and before I had a chance to register further reactions, they were out of sight.

Another half-mile down the road, I was met by a mixed congregation of eighteen mustangs and a score of cows. They were busying themselves in an island of cheatgrass. The meal must have been long awaited, for my approach on foot wasn't taken very seriously. Only when I got within fifty yards of a large black stallion did I realize that being wild means never forgetting your enemies. He kicked up some dirt, made a small circle in front of me, flicked his ears forward, then back, then forward again. He postured evenly on all four feet and snorted. Apparently sensing that I didn't get the message, he pranced forward three or four steps. He shook his head several times and snorted again. He refused to take his eyes off me, and he refused to give ground. I left.

Night upon me, I stopped in an open valley, threw out my sleeping bag on the desert floor, and quickly fell asleep. In the morning, as I crawled out and sat up and looked around the unfamiliar landscape, I saw only two horses, far away on a high hazy bluff. They seemed like wooden sentinels. After I had coffee, I headed back down the canyon I'd come through the night before. I turned back for a final look at the horses. They were nowhere in sight.

It had started to cloud over and drizzle before I got to the Stone Cabin Ranch, and perhaps I would have been well-advised to heed the omen. Joe Clifford and his brother were away working in a gold mine, I was told by his sister who invited me into their bungalow. Joe's wife and two children were huddled in front of a television that sputtered black-and-white, barely discernible stitchings. I had heard that the Cliffords didn't have a telephone, that their mail was picked up once a week, that their power came from a diesel generator, but this. . . .

"Are you from the American Horse Protection Association?" I was asked.

"No."

"Are you from one of those other humane associations?"

"No."

"Are you from a newspaper?"

"No."

"Are you an easterner?"

"No."

"Are you someone new with the BLM?"

"No."

"We've had all we can take of you troublemakers and all these horses," the sister said. "Now we can't do anything. We can't afford to do anything. If we touch those horses the BLM will pull our grazing permit and that will really be the end of our ranch. Not that it matters much anymore." She sighed audibly. The expression in her eyes dulled and she turned away.

As she showed me the door she said, "If you want to see horses, take the road east as you leave the ranch. You'll see plenty of them."

I followed her instructions and soon came upon a broken corral built around a water hole partially hidden by a swelling butte—a great place for a frontier ambush. The Butte Water Hole, as it is known, was a slumbrous pool of grasshopper green. Poles leaned hither and yon, the gate was missing, there were gaping holes, cow and horse dung of all ages littered the arena. The corral was, I guessed, of no particular use to anyone now. Still, I remembered it was the trap where most of the mustangs in the first major BLM roundup in the most populous wild horse state in the Union had been captured.

Sensing that it was either too late in the morning or that I wasn't looking in the right place to see horses, I got back onto Highway 6 and drove east a couple of miles before turning south on an unmarked dirt road. I was following a hunch. On my left was the Kawitch Range, which extends due south for more than fifty miles into the very heart of the Nellis Air Force Bombing and Gunnery Range. In 1962, the departments of Interior and Defense created a three hundred ninety-five thousand acre refuge for wild horses on the bombing range. The sanctuary was established in response to pleas from people around the country who feared for the West's diminishing mustang herds. The bombing range was thought to be ideal because it already had a large horse population and because there were no cattle and therefore little competition for food. At the time of its creation, Interior Secretary Stewart Udall and others thought that the refuge would someday be a national park, and that it could be used for research on wild horses. But because military needs for the base did not diminish as expected, the bombing range was never opened to the public. A pity. From the BLM I learned that the bombing range horse population, which includes those in the reserve, was more than forty-five hundred in 1984. This is the largest single concentration of wild horses anywhere in the West.

A mile down the powdery red road, and still in Stone Cabin Valley, I stopped and stood on top of the cab of my pickup and surveyed the base of the Kawitch with binoculars. Within an angle of thirty degrees

I counted more than two hundred fifty horses on the gentle slopes and bluish-green flats. In months on the road, I had not seen anything quite like this happy crowd of diners. One and all were grinding away at the hand-high grass.

Stone Cabin Valley became famous in 1975 when the Nevada BLM chose it as the first site to round up excess horses after passage of the 1971 wild horse act. By that time, the Nevada BLM had finally become aware that its public lands were being devastated by too many mouths. Range specialists in the BLM, cattlemen, and many conservationists began clamoring for action: they wanted horse populations reduced to their 1971 levels. Stone Cabin Valley was thought to be as good a place as any to begin. A survey showed that almost sixty percent of the valley's range was in poor condition, and it was deteriorating. Several factors were implicated: a succession of dry years, heavy grazing for decades by cattle and horses alike, rancher unwillingness to reduce livestock numbers, and a growing horse population no longer subject to rancher control.

Cows and sheep have grazed in Stone Cabin Valley since the 1880s. The first wild horses in the valley also probably date from this period. One large outfit, the O.K. Reed Ranch in the southern end of Stone Cabin Valley, ran huge herds of horses, and for a while put out stud Thoroughbreds and Morgans. By the early 1900s, ranchers were upgrading the quality of mustang stallions, keeping some of the better ones for cow ponies and selling the rest. In the next several decades, thousands of wild horses were harvested from Stone Cabin Valley. Mustang numbers fluctuated wildly, however, and by the mid-1950s, no more than a hundred horses could be counted. Ten years later, the figure had doubled, and then it almost doubled again in the next half-dozen years. Horses from the Nellis Air Force Base moved in and out of the valley at will, contributing to the problem and the fluctuation in horse numbers. When the BLM proposed to move ahead with a roundup in the summer of 1975, a thousand horses were calling Stone Cabin Valley home. Of these, eight hundred were claimed by Joe and Roy Clifford, making it one of the largest single rancher claim of wild horses on public lands anywhere in the West.

Almost from the inception of the planned BLM roundup, Stone Cabin Valley got an unusual amount of national attention. Thousands of letters poured into the Nevada BLM in support of the gathering, many from people who wanted to adopt one of the mustangs. Conservationists, the Nevada Department of Wildlife, the National Mustang Association, the First Nevada Calvary, and the

National Wild Horse Association all backed the roundup. Wild Horse Annie visited the valley several times and concluded that a roundup was necessary to salvage the damaged range, her only stipulation being that cows should be removed in numbers equal to the horses taken out. The only organization of note which did not support the roundup was the American Horse Protection Association.

The AHPA saw no reason to control the horse population, and it took issue with the ranchers' grazing practices. Apparently believing that the BLM was simply a front for rapacious ranchers, the AHPA filed suit to stop the roundup. In addition, the Washington-based organization took the highly unusual step of sending two graduate students from Humboldt State University into the valley to study the horses. AHPA directors believed that the students' observations would lend support to their claim that there was nothing wrong with large and burgeoning horse herds.

The students were able to count close to eight hundred horses in the valley, and admitted that they didn't get them all. They agreed that the range was in fair condition at best, and they said that some of the horses looked underfed. Contrary to the BLM, however, the students didn't believe that the horse population was likely to continue increasing, and—oddly enough, given their other conclusions—they saw no urgent need to remove them.

In July of 1975, the BLM captured eighty horses in water traps and put them in a holding corral on the Stone Cabin Ranch. But within a matter of days after the roundup, Nevada's State Agricultural Director impounded the horses. He claimed that they belonged to the state and that the federal Wild Horse Act was unconstitutional. Unwilling to keep the mustangs penned up until the issue was resolved, the BLM returned them to the open range. By that fall, the BLM was back at work, and by the following spring a total of four hundred sixty-three wild horses had been removed from the valley.

Large as this gathering may seem, there were still too many mouths for the valley's food supply. Besides the remaining five hundred-plus horses, there were three hundred fifty cows in summer and fall and four times this number in winter and spring in the valley. Sensing disaster, the Cliffords and two other ranchers began voluntarily reducing the size of their cow herds. According to the BLM, one rancher in the valley reduced his grazing use by twenty-five percent "to help improve the range." He intended to return to maximum usage after two years, but never did.

The ranchers became thoroughly frustrated with the BLM. They complained—to no avail—that rambunctious studs were breaking down their fences and injuring saddle horses. They complained—to no avail—that the horses were accounting for almost fifty percent of their water costs. And they complained—also to no avail—that the BLM had told them that the "proper" number of horses for the valley was two hundred twenty-five. Yet by 1980, the figure was more than eleven hundred, and rising. The bureau itself had found that there was forty percent more grazing in Stone Cabin Valley than there should have been. Why, the ranchers wondered, had the BLM allowed Stone Cabin Valley to become the state's equine Bangladesh?

And there was another matter, of less concern to the ranchers and the BLM, but by no means a minor issue to wildlife advocates. In less than a decade, the deer population in Stone Cabin Valley and the neighboring mountains had dropped from fourteen hundred to eight hundred, antelope numbers from two hundred to one hundred fifty. One BLM field report concluded: "With the present level of over-utilization and the expected increase of wild horses over the next few years, it is doubtful whether 'reasonable numbers' of wildlife can ever be achieved in Stone Cabin Valley."

Finally, in the fall of 1983, the BLM sent out a helicopter and a crew of cowboys and rounded up almost eleven hundred horses in Stone Cabin Valley. The ranchers agreed to reduce their herds by nearly twenty percent and keep them at that level for four years. Whether these measures are enough, or have come too late, remains to be seen.

Sixty miles east of Tonopah and two hundred miles north of Las Vegas lies the Fallini Ranch, also known as the Twin Springs Ranch. The Fallinis own grazing rights to 663,000 acres of public land in the Reveille Valley, the southern reaches of Railroad Valley, and a small slice of Stone Cabin Valley. Within the boundaries of this vast operation, roughly the size of Rhode Island, Joe Fallini Jr., his mother Helen, and Joe's wife Sue have deeds for two thousand acres of scattered parcels. In any given year they run between eighteen hundred and twenty-five hundred cows, on what is known as a self-sufficient Hereford, cow-calf operation. Replacement heifers come directly from the calf crop. The optimal number of cows is determined less by forage than by available water. Unlike some ranchers in Nevada, the Fallinis do not share forage with others. This puts them in the seemingly enviable position of being able to control their own destiny. If they damage the range by overgrazing, then they will pay for their mistakes in the long run. If they are prudent with their water and precious grasses, they will be able to profitably perpetuate a century-old way of life.

The Fallini family came to Nevada from Italy over a hundred years ago. Joe's grandfather freighted goods for mines around Eureka until he accumulated enough capital to acquire a small ranch in the southern end of Reveille Valley. Over the years, particularly in the 1930s and '40s, the Fallinis acquired sufficient water and grazing rights to expand the size of the ranch more than ten-fold. Water being the precious commodity that it is throughout the state, and in many areas the critical factor limiting the number of cows, grandfather Fallini and his son Joe reinvested their profits in water development. They made springs more accessible, they put in pipes and ditches and water troughs, they dug wells—a few more than three hundred feet deep—and they installed pumping systems. The list of these privately financed endeavors is impressive: twenty wells, seventeen reservoirs, fifty-seven developed springs, twenty-one hundred feet of tunnels, forty-three miles of pipeline, and forty-nine miles of ditch. The cost over the years: $577,000.

Until the 1971 wild horse act, which forbade ranchers from running horses or rounding them up, the Fallinis' permit allowed them to have one hundred twenty-five horses on the open range. For three-quarters of a century, the family more or less succeeded in confining the free-roaming horse herds to the central and southern portions of Reveille Valley. The Fallinis liked having the horses on the ranch. They enjoyed seeing them and they looked forward to the periodic roundups. Some of the mustangs were given away, some sold for pet food, some were broken and used on the ranch, and some—the bald-face grey ones—were sold for pretty good money. In fact, Joe's father selected for the dapple grey horses with the white faces by putting out good studs with those characteristics.

When the 1971 federal law put an end to the horse breeding side of the ranching business, Joe Fallini Sr. was not pleased. But he tried to convince his son that the government was too big to fight and that it would be in their best interest to get rid of all the horses. "Little Joe," as he is affectionately known by some, thought otherwise. He didn't like the arrangment the BLM had for rounding up the horses, and he didn't figure the horses would prove to be much of a problem. He also apparently didn't believe that their numbers might explode while the BLM stood on the sidelines and watched.

According to the Fallini records, there were one hundred twenty-six wild horses running loose on the ranch in 1971. Then, according to annual counts done from a helicopter by Joe Fallini Jr., the wild horse population on the Twin Springs Ranch increased as follows: 1972—one hundred seventy-eight; 1973—two hundred thirty-eight;

1974—three hundred forty-two; 1975—four hundred sixty-three; 1976—five hundred eighty-four; 1977—seven hundred three; 1978—one thousand ninety-six. By 1984, Joe Fallini was claiming that more than eighteen hundred mustangs were drinking his water and eating his grasses.

Since the mid-1970s, the Fallinis have been edgy, mad, and preoccupied. Because of the unusually large wild horse population—the largest found on any ranch anywhere in the nation—the costs of operating their spread have increased dramatically. The Fallinis blame the BLM. For a decade they have demanded that the bureau remove all horses above the number on their ranch in 1971, citing the Wild Horse and Burro Act that provides for "protection, management and control of wild free-roaming horses and burros on public lands." All that they feel they got for their efforts were excuses, delays, and periodic reports to the effect that the BLM doesn't have enough money, or has higher priorities, or that it has decided to increase the number of wild horses deemed optimal for their allotment. By the late 1970s, the Fallinis were so angry that they told their attorneys to do whatever was necessary to get the horses removed. But legal maneuvers take time, and Joe Fallini, an energetic, impatient, and determined sort who has lived just about all his life on the Twin Springs Ranch, felt that time was working against him. In 1983, while public lands chairman of the Nevada Cattleman's Association, he took his case and that of other Nevada ranchers to a U.S. Congressional hearing on wild horses. He demanded that the BLM be forced to do its mandated job, and that new laws be passed that would facilite the disposal of excess wild horses. He also made a strong case for ranchers like himself who, he said, were on their way to bankruptcy because of the proliferating mustangs.

Helen Fallini: "The way we used to do it was if the range was good we let the cattle build up a bit. If the range was bad, we sold them down. It just made good business sense, because if you're gonna be in business you've got to take care of your range."

Nineteen eighty-one BLM Report on the Twin Springs Ranch: "The past management effort put forth by the Fallini operation has been very good and has maintained good overall range condition."

Helen Fallini: "My husband didn't want to see all the horses go. And I didn't want that either. But we made one hell of a mistake by not getting rid of all of them in 1971. Some of the ranchers that were in our position just plain shot them. They could see the handwriting on the wall when that law came in and the BLM came with it."

Nineteen eighty-one BLM Report on the Twin Springs Ranch: "The wild horse population on the Fallini property is interrupting and altering the rest-rotation system which has operated for years. In several locations where both cattle and wild horses use the water it has been necessary to move the cattle out early to prevent overutilization. In the past, the Fallinis have pumped water at their own expense to supply water for wild horses."

Joe Fallini Jr.: "The horses are eating and drinking us out of business. Those horses have to go."

Milt Frei, the BLM's chief wild horse expert in Nevada: "There are areas worse than the Fallinis, plenty of areas that are worse off."

Nineteen eighty-one BLM Report on the Twin Springs Ranch: "Cattle use in critical areas can and has been reduced by using available water to rotate the cattle, although some use will still occur in these areas. Wild horses cannot be controlled, particularly when populations are large. Therefore, some areas of the range receive uncontrollable grazing use year-long. The increasing wild horse populations and herd use areas, the overall use of some areas, the alteration of the rotation system, and the potential deterioration of the rangeland resource can be controlled by controlling wild horse populations at reasonable numbers. Wild horse numbers must be maintained at a level which will maintain the ecological balance of the resources."

Helen Fallini: "They should remove those horses. We pay for using that land, more than a dollar and a half a month for each cow we have out there. And then we have all kinds of other expenses on top of it. It's expensive, real expensive, running a ranch."

Paul Jancar, BLM wild horse expert in Nevada's Winnemucca office: "By the time you consider range improvements made by the BLM, the rancher is paying less than fifty cents a month to keep a cow on public land."

Helen Fallini: "The problem is those eastern wild horse groups come out here and look around a little bit and then they go back to the schools and have third- and fourth-grade kids write about these horses. The teachers tell them what to say and then their vote goes to their congressmen and it counts as much as our vote. All those votes go in, more than we got, and we have to live with them. It's pretty discouraging because it would be just like me going east and complaining to those people about their houses on the beach. It would be like saying we want those houses removed because we want the beach looking like wilderness areas and parks. They wouldn't like that and we don't like them coming out here. There's no difference in the world."

A year after I first heard about the Fallinis and their frustrations with wild horses, Joe invited me to come to his ranch to appreciate their problem firsthand. As it turned out, I was introduced to his horse problem even before meeting him. At about six o'clock on a hot June evening I was driving north on Highway 375, the southern end of Railroad Valley, in the southeastern corner of his ranch, when off to my left, near the base of the Reveille Range, I saw a great gathering of animals around a water tank. Reasonably certain that I wasn't looking at cows, I took the first dirt road and headed toward them. A little more than a mile west of the highway, I came upon a large sturdy corral, a huge blue water tank that looked like a grain elevator, and an unsettled gathering of wild horses, by quick count, one hundred forty-six in all.

Parking beside the thumping gasoline engine that drives the well pump, I got out and began walking toward one of the open corral gates. Dozens of the horses began a slow retreat toward the bleeding flanks of the Reveille Range. One stallion, five mares, and two colts left at a brisk gallop in the direction of the highway. But better than eighty of them, mostly bays, dark sorrels, and grey roans, refused to be intimidated by my sudden arrival. They formed a three-quarter circle around me at about seventy-five yards. If I moved a half dozen steps in their direction, those closest to me retreated eight or ten steps. If I turned back toward the corral or the truck, they suddenly stopped. Some whinnied and snorted and tossed their heads, others violently shook their tails. Harem stallions checked to see that their mares and young were safely in the rear. A couple of the larger, thick-necked ones—the biggest stallions—regained ground they had given up, and then added to it. For what seemed like long minutes an arc of observant eyes stared at me. I read insolence in their bony faces.

My adrenal glands working overtime, tossing all caution and common sense to the wind, I began stalking small bunches of mustangs. At one point, I pursued three bachelor studs—a blue roan and two dark bays—about three hundred yards back toward the highway, into the treeless grey-green desert. For better than ten minutes I gained on them, then lost half the distance, regained ground, then lost even more ground. They faced me head-on, they turned to one side or the other, they pawed the sage and the bitterbrush and the dusty earth, they crowded one another and then backed away, they snorted and shook their heads. They were, I think, trying to tell me that I'd come far enough and ought to mind my own business. Perhaps they were saying that they had decided beyond all reasonable doubt that their intense desires for companionship and elusive mates could wait. Finally, the two bays scampered away from the blue roan and ran a hundred yards or so before stopping to reappraise the situation.

After I'd taken several pictures of the proud roan, he also decided that he'd had enough of me. He circled, took off in a gallop, then walked slowly toward the sloping mountain. By this time, I'd forgotten about his traveling buddies and began walking back to the corral. Suddenly, for no particular reason, I looked back and immediately felt that indescribable rush called panic. The two bay bachelors whom I had forgotten about were fifty to sixty yards away and galloping straight for me. Not knowing what to do, certain only that I couldn't outrun them, I raised my long-eyed Minolta and took a couple of steps directly toward them. Whatever they thought they saw, they stopped dead in their tracks—and I got the hell out of there.

By the time I returned to the safety of the corral and my pickup, the horses had begun a slow retreat to higher ground. By now, the bright light of the long summer day was dimming. Soon, I lost sight of all but a few of the mustangs. They had disappeared behind juniper and pinyon and the safety of ridges, into the darkening shadows of their mountain home.

Not long after my arrival at the Fallini ranch, Joe and his wife Sue took me for a heaving, gut-wrenching ride into Locks Pass Basin and Cold Cut Basin, narrow, tributary valleys a couple of miles east of ranch headquarters. By Joe's estimate, there were one hundred sixty mustangs in the basins. Before the three of us stopped counting horses on the flats and the alluvial slopes and the sharp ridge tops, we had passed one hundred twenty. The number would have gone higher if the horses hadn't started to flee the moment they saw dust and heard the roar of the truck.

On the return trip to the ranch, Joe said little, other than to remark that there was enough feed left in the basins for about three weeks. Feed for the horses, that is. Joe wouldn't be pushing cows into the basins anytime soon. He sighed heavily and screwed up his mouth and looked over at Sue, who let loose with some unflattering words about the BLM. I didn't have to be told that the sight of so many horses had put both of them in a bad mood. By this time, though, I had learned that Joe Fallini wasn't one to run from adversity.

According to a newspaper article that I'd seen on a restaurant bulletin board a couple of days earlier (later confirmed and elaborated on by Sue), in 1980 two motorcyclists from the east were passing through the unfenced Fallini ranch when they decided to see how much fun it was to chase cows. As the twosome enjoyed themselves chewing up grass and frightening and scattering the cows every which way, Joe just happened to be driving up the highway in a stock truck. He pulled over to the side of the road and waited until the pranksters stopped and got off their bikes. Then he hit the gas and slowed down only enough to make sure that one pass over the bikes would render them useless. When the disconsolate pair tried to file charges against Joe, the sheriff told them to forget it. He advised them to pick up their wrecks and get lost.

Another day, Joe and I got in his battered, red High Sierra pickup and headed south into Railroad Valley, northeast of the well where I'd first seen so many horses on his ranch. He wanted to check on well pumps. If the engines that run them break or are out of gas, then there won't be water for his cows. Since the horse population has become so large, Joe has had to check his pumps more often than he'd like to, a hell of a lot more often than he'd like to, he says. In 1971, he could refuel the engines and check to see that pipes were in good order and troughs were full of water once every four days. Now he was having to do the same job every other day, sometimes every day. With so many mustangs on the ranch, Joe can never be certain which well the horses will decide to take over, or what damage they'll do to pipes and troughs if they conclude that they're not getting enough to drink.

At one well, Joe said, "This one is pretty high in sodium, so the horses won't drink the water. That's why the feed is good around here." At another, he exclaimed, "Right here we're producing water for about three hundred horses, and we're watering seventy-seven head of cows. That's exactly how many cows we got here."

It didn't take long to be convinced that wild horses were on the Twin Springs Ranch in abundance. In a little more than two hours, at a time of the day when one expects to find the horses languishing in mountain shade, we saw no less than five hundred of them. And that was without binoculars. With each herd that ran from us as we approached a well, Joe got a little madder. The horses, he said, were drinking all his water and eating all the best grasses and wreaking havoc with his bank account. He cussed to himself and he cussed at the BLM; at one point, he seemed to be cussing at the Fates. Finally, as if to signal the depths of his predicament, he looked over at me and pulled hard on the bill of his orange-and-brown cap, with the oval label that read "Max Rouse, Auctioneer."

I was stunned by the number of horses I had seen. In the previous year, I had heard so many people bad-mouth ranchers that when I came to see the Fallinis, I was prepared to believe that they, too, might be in the business of calling cows horses in order to exaggerate their "wild horse problem."

As Joe and I drove on to another well and saw scores and scores of mustangs kick up dust and take flight at our approach, it occurred to me that this might be as close as I'd ever get to seeing a real-life picture of the immense mustang herds that just about owned Texas in the eighteenth and nineteenth century. In 1777, for example, one Franciscan missionary described them as "so abundant that their trails make the country, utterly uninhabited by people, look as if it were the most populated in the world. All the grass on the vast ranges has been consumed by them, especially around the waterings." Estimates of some horse herds at this time ran as high as three thousand. Seventy-five years later, on the prairies between Corpus Christi and the Rio Grande, travelers were claiming they saw tens of thousands of wild horses, some over fifteen hands and weighing one thousand to eleven hundred pounds. When Ulysses S. Grant and a column of soldiers crossed the sun-baked plains of southern Texas in 1846, he was awestruck at the sight of wild horses. He was later to write, "As far as our eye could reach, the herd extended. There was no estimating the animals in it." So numerous were wild horses in Texas that map makers referred to the great open spaces between the Rio Grande and the Nueces River as the "Mustang Desert." On a map of Texas, Stephen F. Austin scribbled: "Immense Herd of Wild Horses." J. Frank Dobie, a student of the history of Texas's wild horses, guessed that by 1850, when mustang numbers probably reached their peak in the West—roughly two million—Texas had half of them.

The Fallinis' grazing rights—the number of cattle they are permitted to run—are based on the ownership and control of water. With each new water development, the BLM allows so many additional cattle within a four-mile radius of the well head. Which explains why the family has poured so much money into wells and pipes, and why Joe can't get the horses off his mind. In 1984, Joe estimated that the horses on his ranch were consuming more than five million gallons a year, better than fifty percent of everything pumped. A range cow drinks five to ten gallons of water per day, a wild horse two to three times this much.

The horses, Joe said, were costing him at least $50,000 a year. Most of these expenses, which he can detail with the scrupulousness of a squint-eyed cost accountant, directly relate to water: extra gasoline for the well pumps, damage to wells and pumps and springs and troughs, and the wages of an extra hand to inspect the wells and do repair work. At one spring Joe had to bring in an excavator and a Caterpillar™ to put in new pipe and rebuild a wooden tunnel. It was then destroyed a second time. His total bill for the two efforts: $30,000. In 1977, Joe had a different kind of problem with water; several mustangs drowned in his troughs. Crazy from locoweed poisoning, they had lost the ability to back away from the water. To prevent pollution of the water holes, Joe had to drive almost a hundred miles a day for several months, checking for carcasses.

There's no easy or humane way to control horses' drinking habits, or how much they'll eat before looking for better pastures. The rest–rotation system used by the Twin Springs Ranch, where wells are shut off to force the cows to move from one pasture to another and thereby prevent overgrazing, doesn't work very well when you've got lots of horses to contend with. Helen Fallini explains why: "The minute you close a water hole down, the horses just move to the next one, always ahead of the cows. I've seen the horses go eighteen to twenty miles for water in a single day. We have to close down a lot of them to try to keep care of that range. If we don't there's no feed to put the cows on. But the worst part of it is that you can close down one well here and your cows will move out, but the horses will move down to the next water hole faster than the cows, get a good drink, and then double back and take the feed we're trying to protect. That's where our problem is." Joe repeats his mother's words, and then adds that the horses have done so much damage that he can no longer use the rest–rotation system that he grew up with and had hammered into him as a student at the University of Nevada.

Joe Fallini, it so happens, is not your average western rancher, at least when it comes to resting rented rangeland. Nor does he think like most range experts who have worked for the BLM. It seems that it took the BLM until 1968 to decide that continuous grazing of forage throughout the growing season was an unacceptable practice. Yet five years later, a report from the Department of the Interior found that almost fifteen hundred of seventeen hundred sixty grazing allotments in six western states were being grazed continuously by livestock. As late as 1975, "most of the forty-seven million acres of public rangelands in Nevada were subjected to continuous grazing."

At one well, about fifteen miles south of the ranch, the corral looked as though it had been slammed several times with a wrecking ball. Before I could ask what had happened, Joe was drawing my attention to a helter-skelter pile of timbers and tangled barbed wire. He barked, "This kind of crap is uncalled for. We seldom worked on fences before the horses got in here. Now we're working on them all the time. It wouldn't happen if the BLM would follow the law and take out these horses." I had noticed that some of the corrals were reinforced with sturdy guardrails. Joe explained that reinforcing the corrals with guardrails was about the only surefire solution that he'd found to slowing down costly horseplay.

The horses, Joe notes, have neither manners nor concern for others. They love nothing better than to play games with his cows during a roundup. He can tick off a dozen instances in which mustangs have mischievously scattered a hundred cows on a roundup march, adding hours to the strenuous annual task. The horses have gotten so bold that they've even come right down to the ranch house to pick fights with his saddle horses or steal a copulation with one of his mares. They have threatened his young, ranch-savvy daughters, so that Joe and Sue can no longer allow Anna and Rindy and Corrina to ride or play on the open range by themselves.

Joe's mother has a story she likes to tell about the horses' uncouth ways. "We had this girl working for us and she was out riding and this wild stud attacked her. Then he did it a second time. Joe was there in his pickup and he got between them three times before he finally shot the horse. Then he had to go into the BLM and report it, and the way they acted you would have thought he shot the King of England."

While Joe and I went from one well to another, he gave me lessons on the connections between this or that soil type and the varied kinds of plant life we were driving through: shadscale, winterfat, fourwing saltbrush, galleta grass, needle-and-thread grass, squirreltail, and lots of others. Joe wanted to impress on me that there're no easy generalizations about what kinds of grasses you will find on the seemingly seamless desert floor. The horses, or too many cows—or both—are to blame for successional changes, but an awful lot depends on soil: whether it's hard, or sandy, or shallow, or gravelly. And soil type, Joe contends, can change as quickly as the front wheel of his pickup can lurch from one badger hole to another.

At one point, we stopped, got out of the truck, and stepped into a cobwebbed blanket of halogeton. Joe pulled up a hunk of the wiry, shiny green invader and explained that the plant uses valuable water. Cows don't often eat halogeton, but when they do, it makes their bones brittle. Which means expensive losses if their bones break while you're giving them a good push on a roundup. Once halogeton gets a foothold, it is extremely difficult to eradicate. In 1954, the U.S. Congress passed the *Halogeton glomeratus* Act to fight the spread of the weed on western rangelands. The effort and the money spent were for naught. A single plant can produce forty thousand seeds. When the plant dies, the seeds blow all over the range; they will germinate for up to ten years after being produced.

I asked Joe about the significance of the striking wild mustard that filled the middle distance.

"When you see it, it means there's been too much grazing."

"How much is too much?"

"This valley now has one-tenth the carrying capacity it had in 1971."

Was he exaggerating? I had no way of knowing. But I was convinced that wild horse activists, who can't find the time or won't accept Joe's invitations for a tour of the ranch, don't know either. I doubted that the BLM could settle my doubts.

Joe got down on his haunches, pulled back a clump of sagebrush, and pointed to some grass that he said horses hadn't eaten because it was too hard to reach. He went on to say that what makes horses so destructive is that unlike cows, they nip grasses and forbes close to the ground, often killing them. As we stood up, I was reminded of a saying I'd heard from a cowboy several weeks before. He said, "Put a cow in a pasture until it gets skinny. Then put a sheep in the same pasture until it gets skinny. Then put a horse in that pasture and it'll get fat." This buckarooing wisdom seems to have some basis in scientific fact. A study at Colorado State University found that mares graze plants more closely to the ground than cows, and that horses are more likely to have a severe impact on range resources than cattle when "proper stocking rates" are exceeded. The National Academy of Sciences agrees, noting that horses crop forage more closely because, unlike cows, they have both upper and lower incisors.

Joe showed me lots of galleta grass. I came to know it well: it seems to be everywhere on the ranch. Galleta grass is common where desirable plants have been eliminated by overgrazing. It is low in carotene and protein and phosphorus, and is less nutritious than other grasses, especially that gastronomic delight, Indian ricegrass.

Handing me a small bunch of the lacy-white grass, Joe explained that it grows well on hot, slightly alkaline soils. Indian ricegrass is everyone's preferred perennial. Cows like it, horses love it, even the Indians favored it. They boiled it, they ground it, they fried it, and they mixed it with meat. Highly palatable throughout the year, the grass is a sensitive indicator of range conditions. When in short supply, it usually means that the range is deteriorating. In valleys east of the Twin Springs Ranch, it has been shown that heavy grazing on

Indian ricegrass over several years results in its virtual disappearance.

Joe drew my attention to the slippery slopes of Quinn Canyon Range, which defines the eastern edge of Railroad Valley. "See all that red?" he said. "It's all cheatgrass now. If you came out here before '75 when the horses first came into this valley, that same area was full of ricegrass. Now the ricegrass is all but gone and it really pisses me off because there's nothing I can do. Goddamn, you work your ass off and those goddamn BLM bureaucrats screw everything up for you."

What's wrong with cheatgrass? One problem is that it ripens early in the season and all at once, which means there is feed for only part of the year. Mule deer eat tender cheatgrass, but cannot overwinter on it. The plant remnants are sparse, and this increases the likelihood of soil erosion. The yields of cheatgrass vary enormously from year to year, and it is not as nutritious as many native bunchgrasses. Though once highly touted by stockmen, and the dominant species on several million acres of western rangelands by 1950, cheatgrass has the further liability of being much more flammable than other grasses. Several hundred times more flammable, in fact. This extends the fire season, and fires that begin in cheatgrass then spread into areas that otherwise would not be affected. Cheatgrass seed can survive a prolonged drought, then suddenly germinate within a couple of days after a rain. Unlike native grasses, its seed production and germination rates are good, and it has a great competitive edge over native grasses. Cheatgrass can withstand repeated trampling and grazing. Frequent disturbance by large herbivores selects for rapid germination and growth. Intermountain bunchgrasses die when heavily trampled by cows and horses.

Cheatgrass probably evolved in association with horses and camels on the Eurasian steppes. It first appeared in the United States in the 1860s, and may have reached Nevada as early as the turn of the century. Sometimes known as "military grass," it could have been brought into the state by military trains carrying calvary horses. When the cars were pulled off onto sidings to be cleaned out, the cheatgrass seeds were scattered. Although known as an "invader" by biologists, it has trouble getting established where there are healthy stands of native grasses. On the other hand, cheatgrass colonizes with amazing speed where ranges are overgrazed.

Joe said that in Reveille Valley, to the west, the horses had taken such a toll that areas formerly covered with Indian ricegrass had turned to cheatgrass and lots of rabbitbrush. "And that stuff—rabbitbrush—is worthless." So worthless that he had to reduce his cow herds in Reveille Valley to almost nothing, just enough to maintain his water rights under state law. He's hoping that a long rest will restore once-common grasses, allow him to once again run hundreds of cows just as he had before mustangs turned the landscape penurious.

"How much do the horses eat a day?" I asked him.

"Probably thirty pounds. But they're very selective, much more than my cows. Particularly in the spring when high-quality forage insures a good fecundity rate and healthy foals."

Joe doesn't really know just how much similarity there is between the diet of his cows and that of the public's wild horses. But he couldn't be more certain that there's plenty of dietary overlap. Joe gets downright furious when he has to listen to someone like J. Stewart White, president of the Nevada Humane Society, who will side with a young girl's pronouncement to the effect that, "Sheep and cattle eat all the grazing up and the horses get all the blame."

One of the few reliable, in-depth studies on the question of dietary preferences was done by a group of range scientists from the University of Wyoming in the late 1970s and early 1980s. They found that cow and horse tastes overlap considerably, from a low of fifty-two percent in early summer to a high of ninety percent in winter. Sixty percent of the winter horse diet is composed of grasses and sedges; the comparable figure for cows is sixty-six percent. In addition, horses and cows have almost identical preferences for such shrubs as winterfat, shadscale, and fourwing saltbrush. In areas densely covered with both horses and cows, these figures are even higher. On the other hand, direct fighting over food is uncommon between horses and cows. Horses spend the majority of the daylight hours in the summer and winter grazing and foraging. In the summer, cattle prefer eating in the morning or late afternoon, but will eat all day long during the winter. Both will eat at night throughout the year. Horses don't like cows around when they're drinking, though, and the Wyoming study found that they bite and kick cattle to keep them away.

Joe likes to keep a two-year reserve of feed on the range. He reminded me that widely fluctuating rainfall characteristic of all desert climes necessitates more than a casual sense of the future. The Twin Springs Ranch averages seven inches of rain a year. But one good year of eight or ten inches can be followed by a couple of years of only two or three inches. Some ranchers try to protect themselves by growing their own hay. For the Fallinis, this is a problem; they don't own enough of the right kind of land to grow a good alfalfa crop. When Joe and I talked, he didn't say exactly how much reserve he thought he had left, but it was less than a two-year supply—and that could prove disastrous. By BLM estimates, probably conservative, his cows and the public's horses and wildlife were overutilizing forage on the Twin Springs ranch by some forty percent.

On the return trip to the ranch, Joe said that I ought to go up to Little Echo Canyon in the Reveille Range, which is a winter and summer home for mule deer and antelope. He said that in the early 1970s the canyon had a large mule deer population; browse was plentiful, and so was Indian ricegrass. "Then in 1973 the destruction by horses began." Little Echo Canyon became a "nightmare." The horses ate grass right down to the roots, something his cows wouldn't do even in the worst of times. Joe said that the horses had nearly destroyed the inlet to Little Echo Spring, which was the one important watering place in the canyon for deer. Several days later I took a trip to Little Echo Canyon by myself. From what I could tell, Joe hadn't exaggerated. There was an abundance of stud piles around the available water, and there was no sign at all that deer had been in the area recently. I did see some cheatgrass and other inferior types. Mostly I saw lots of bare ground. I didn't, of course, know firsthand what the canyon had once looked like. But then it didn't take much imagination based on what I'd heard and read.

According to a BLM resource report on the Twin Springs Ranch, while water is the single most important factor determining wildlife numbers, forage is also critical. Spring "green-up" provides necessary nutrients for mule deer growth in weight, body size, antler development, fawning, and nursing. Livestock and wild horses have similar nutritional needs, and they feed intensively on new spring growth. Winter ranges, which provide browse (high in protein), are relatively scarce on the Fallini ranch. The available winter range is in fair to poor condition. It is needed and used by deer and livestock as well as wild horses.

By BLM estimates, the deer population on the Twin Springs Ranch dropped from eight hundred sixty in 1970 to four hundred ninety by the early 1980s, antelope from one hundred seventy to one hundred thirty-five. These reduced wildlife numbers are well below what the bureau considers "reasonable" for the ranch.

The BLM and ranchers may be on congenial terms in certain parts of the West and in the minds of many outsiders, but you'd never know it by talking to the Fallinis. Pure and simple, the Fallinis seem to see the BLM as the enemy.

Take the issue of horses on the Twin Springs Ranch in 1971. Joe is positive that he had one hundred twenty-six on his spread at that time, and that he knew precisely where they were. The BLM doubts this figure; it claims that the real number was two hundred ninety. When Joe has tried to see BLM records showing how the agency arrived at its figure, he has been told that "the records just cannot be found." Joe Fallini is equally stupified by the BLM's continually changing idea of the optimal number of wild horses for his ranch. In 1971, he claims, the figure was judged to be eighty to one hundred. In subsequent years, the number changed five times—always upwards. At last reckoning, the BLM thought that three hundred fifteen mustangs would be ideal for the Fallini Ranch.

Whatever the real number of mustangs on the Fallini operation in 1971, all questions could have been resolved two years later when, by law, ranchers had the right to "claim" horses on the open range. In theory, a rancher could have said that every free-roaming mustang on his allotment belonged to him. And some ranchers did precisely this. Mike Mitchell, the BLM cowboy in Battle Mountain who has worked with the Fallinis over the years on the wild horse issue, asserts that efforts were made by the BLM to get ranchers, including the Fallinis, "to see the light." But, according to Mitchell, his pleadings went unheeded. "Ranchers in this part of the state," he says, "are a pretty traditional lot, out of touch with modern progressive ways."

Or was something else on their minds? In the 1940s the Grazing Service, forerunner of the BLM, wanted to get rid of Nevada's wild horses, but cattle ranchers wouldn't go along with the idea. Some feared that once the horses were removed, they would be replaced with sheep or wildlife. Others simply couldn't envision a positive payoff. While the government was preaching, "A horse will eat twice as much as a cow," ranchers were countering with, "But the horse can live where a cow will die." Still other ranchers simply didn't want the government tampering with the secure income they were getting from the periodic sale of the mustangs, and with the envisioned prospect that growing urban centers in California would increase demand and raise prices.

Joe Fallini Jr. will have none of Mike Mitchell's argument. He says that the BLM district office in Battle Mountain gave ranchers the right to claim their own horses all right, but if they did so they would be charged so much for the BLM-conducted roundups that they'd lose money. And, worst of all, the BLM would give them demerits on their grazing records, which, if one accumulates enough of them, are a sufficient reason for the BLM to revoke a rancher's permit to use public lands. Rather than lose money and risk trespass actions, Joe and his father did nothing.

Then there's the issue of the BLM's desire to convert the Fallini ranch into a wild horse preserve. When the preserve proposal is mentioned, Joe's eyes turn strychnine pink. In 1976, a BLM district manager proposed acquiring all private lands and water from the Fallinis and neighboring ranchers. The plan called for having nearly two thousand horses on the Twin Springs Ranch and on others in Stone Cabin Valley. Once the preserve was firmly established, the BLM—a BLM document states—would then eliminate all wild horse herds in the rest of the state. In the words of Roger McCormack, Nevada's Associate Director for the BLM, "the plan was a good one because the horse herds on the Fallini ranch were a geographical extension of those on the Nellis Wild Horse Range, and because Joe Fallini was a lone operator." If the idea was to establish a wild horse preserve for the benefit of the public, few choices would seem to be worse. The Twin Springs Ranch is one of the more out-of-the-way places in Nevada.

The proposal went nowhere. But never mind: Joe sees it as just one example of a BLM conspiracy to get rid of him. If the BLM can't do it by buying him out, or by indefinite postponements in taking off large numbers of horses from his ranch, then he believes that the bureau will do so by means of harassments and fines and trespasses. In 1976, the BLM forced the Fallinis to conduct a special roundup to ear-tag their cows. The purpose of the tags was to make it easy for the bureau to keep track of how many were on the range. Subsequently, Joe discovered that the BLM made little effort to use the tags.

In that same year, the Twin Springs Ranch was charged with running six hundred more cows on the range than permitted. Joe and his father agreed that they were above their legal limit, but said that this was because the BLM had counted their cows just before the annual roundup. If they had been counted shortly after the roundup, when yearlings and fat cows had been shipped to market, the bureau would have gotten a figure considerably lower than that allowed by law. The Fallinis were incensed because BLM policy has no flexibility, even more so because they considered the bureau's figure for cows in trespass to be "out of line with reality." When the matter was finally settled, the BLM and the Fallinis reached a compromise. The Fallinis paid a trespass fine for one hundred twenty-five cows.

Joe tried to persuade the BLM to let him remove mustangs from his ranch at his own expense. He said he was willing to catch them in his corrals and turn them over to the BLM. All the agency had to do was bring in trucks and take them away. The BLM balked at Joe's removal plan. It feared that this would establish a dangerous precedent, that other ranchers would soon be lining up to get rid of wild horses on their allotments.

In 1980, the BLM removed four hundred sixty horses from the Twin Springs Ranch. But this didn't mollify Joe Fallini. He felt that this was not nearly enough, that the BLM is content with a horse population several times greater than is justified by the act of 1971. The reason, it seems, has almost nothing to do with the agency's assessment of damage done by the horses. District managers and range experts openly admit that the Twin Springs Ranch has an acute problem. Jim Fox, the BLM district manager in Battle Mountain, says that the Fallinis have been "good managers" of their public allotment, and that this has worked against them. If Joe ran a sloppy operation and gave clear evidence of overgrazing, then the BLM, Fox says, would have done mustang roundups on his ranch some years ago. Another BLM employee has said that the bureau's slow-footed intransigence is rooted in fear of wild horse activists: of their lawsuits, and of the negative publicity they can generate.

Finally, Joe lost interest in discussing the horse problem with the BLM. By the early 1980s, he had gone to his lawyers and told them not to settle for anything less than the same number of wild horses that were on his ranch in 1971—one hundred twenty-six by his reckoning. Furthermore, he wants the horses confined to the same area where they were at that time. When asked to back up his demand, he cites the wild horse law: "Management of wild horses and burros must be limited to areas of the public lands where herds existed prior to 1971." Joe also has maps indicating the location of mustangs in 1971. The BLM, it seems, has nothing more than hunches. When I asked Jim Fox if he could identify the 1971 horse area on the Twin Springs Ranch, he said, "It'll be tough."

"I'm going to make it hard on the horses if something serious isn't done soon," Joe kept telling the BLM. And Joe meant it. In the fall of 1983, he called the BLM district manager and informed him that on October 21 he was "going to cease to provide water for the horses." Joe didn't tell the BLM what he had in mind—guardrails four to five feet off the ground around the perimeter of his water holes. They wouldn't keep out all the horses, but they'd encourage plenty of them to look elsewhere for a drink.

Initially, the BLM apparently paid little attention to Joe's threat, or the water needs of the horses. According to Jeff Rawson, of the Battle Mountain BLM, Joe put the barriers up as promised. They were still there in January of 1984 when the BLM did a helicopter census of wild horses on the Fallini ranch and came up with a figure close to sixteen hundred. Shortly thereafter, during a twenty-day period in late January and early February, the BLM rounded up four hundred horses. The bureau would have removed another hundred if the herds they were chasing hadn't been thoroughly scattered.

By the spring of 1984, the BLM got a court order to force Joe Fallini to remove the guardrails. Joe promptly removed all but one of them. He figured that blatant defiance would force the court to decide whether the water belonged to his cows, or to horses over which he had no control.

Does Joe Fallini, or any Nevada rancher, have the right to prevent federally managed wild horses from drinking water that he has developed at his own expense? By Nevada statute, wildlife has a first claim on water. But in Nevada, wild horses are not considered wildlife. Indeed, at present, wild horses may be third in line at the spigot after wildlife and the water needs of ranchers. There's nothing wrong with being third if there is plenty of water to go around. But in much of Nevada, scarcity prevails. If, in a test case, the courts were to decide that cows do in fact have a prior claim on water, then the remaining available water could determine horse numbers.

Joe Fallini is mighty close to raising an even larger issue in the state, one with which the courts have not yet dealt. Is water really any different than real estate? Because most water in Nevada is on private property, and because the majority of the West's wild horses are found in Nevada, this issue has profound consequences. The Sweetwater Ranch, a hundred or so miles west of the Fallini operation, could prove to be the test case in this issue. The Sweetwater Ranch, south of Hawthorne, has title to seven thousand acres and a BLM permit to run cows on 234,000 acres of public land. Based on bureau estimates, there were roughly 180 wild horses on the ranch in 1971. By mid-1984, the figure was close to six hundred, more than three times the number considered optimal by the BLM. At that time, the projected date for bureau roundups: "the unknown future." The reason: "prior commitments to make removals of horses on other ranches."

The Sweetwater Ranch runs eight hundred cows on the range from November to the middle of April. On a twelve-month calculation, this is some eighty fewer cattle than permitted by law. "The number of cows out there," Julian Smith, the ranch's lawyer, says, "is below that allowed because the owners don't want to see horses starving to death." He adds that this might well have happened if it hadn't been for an unusually good string of winters from 1982 to 1984, what he likes to refer to as "two hundred percent years." Benny Romero, who has managed the Sweetwater Ranch for ten years, agrees. Romero says that he has seen water where no one has seen it before. Romero wonders what would happen to the horses in a "normal year," or a "bad year"—a predictably recurrent event in unpredictable desert climes.

Smith points out that water in Nevada belongs to the state. "So the U.S. government has no right to use the state's water and therefore the wild horses, which are federal property, have no right to it." Ranchers get permits for what is known as "consumptive use" of water for their cows. Smith says that should a court determine that water is no different than real property, and should the owners of the Sweetwater Ranch get a permit to use all of the water on the ranch—with the exception of that needed by wildlife—then the BLM would be forced to remove all of the horses.

"But would your clients want all the horses removed?" I asked Smith.

He said that they would be happy to have one hundred eighty or two hundred mustangs on the range, the number there in 1971. "They're horse lovers beyond customary measures. They raise quarter horses and they train them. Their children ride them. They never thought of complaining about the horses until their numbers got to be too much."

"They tried working with the BLM?"

"They asked, then demanded to be put on a priority list for round-ups. Nothing worked. They only wanted to file a lawsuit as a last resort."

It looks as though it won't be necessary to take the government to court. In February 1984, the BLM removed three hundred eighty mustangs from the ranch.

On October 4, 1984, a United States district court in Las Vegas ruled in favor of the Fallinis and against the defendants: James Watt, Robert Burford, James Fox, and Edward Spang, the state director of the Nevada BLM. The judge said that the Fallinis had established, to the satisfaction of the court, "damage to fences, forage, water holes and other property . . . and they have spent thousands of dollars in efforts to reverse the destruction caused by the free-roaming and marauding wild horses." He went on to say that for years ranchers have been required to keep their herds from straying onto public lands which they did not rent, and "therefore it is unfair to hold the federal government to a less stringent standard than that imposed on private ranchers." In 1978 Congress had amended the Wild Horse and Burro Act, providing that it is the duty of the Secretary of Interior to keep inventories of wild horses, to determine when there are too many, and to "immediately remove excess animals from the range so as to achieve appropriate management levels and to protect the range from the deterioration associated with overpopulation." In the case of the Fallinis, particularly in light of their frequent appeal to the BLM for removal of excess horses, the judge said that, "This neglect of duty by the BLM is unconscionable."

The judge went much further than merely requiring the bureau to remove excess wild horses or those that were using the ranch's private property. He said that he was "enjoining the defendants from allowing future intrusions by the animals. The act does not explicitly require the defendants to keep the animals from entering private lands. Nonetheless, in light of the history of public lands management and the evidence offered by plaintiffs describing the damage continually being done to private lands, it seems reasonable to grant plaintiffs' request."

The implications of the court's decision are profound. Will the BLM have to put a fence around the mustangs to keep them off the Fallinis scattered parcels of unfenced private land—two thousand acres spread over 665,000 acres? Or will the bureau have to remove all horses from the mountains and valley bottoms of the Twin Springs Ranch just to abide by the letter of the court's ruling? And what does the judgment portend for other herds of wild horses?

In a footnote to the court's decision, the judge took off his robe and really took the BLM by its beard. He said that, "In studying the pending motions, the briefs and documents and exhibits in this case, one is utterly appalled by the almost criminal neglect by the BLM of its statutory duties to reduce excessive herds of wild horses in Nevada. Cattle ranchers have been and are being abused by the deliberate inaction of the BLM. At all levels within the Interior Department we see gross neglect, misfeasance and malfeasance. The agency and even its lawyers are experts at delay, obfuscation and plain old bureaucratic stonewalling.

"Surely and certainly the plaintiffs and many other Nevada cattle ranchers are being driven out of the business so important to the economy of Nevada. Complaining ranchers are intimidated and abused by the BLM through its control of grazing upon the public lands. The BLM holds a life and death hammer over the ranchers' heads.

"What a Frankenstein monster has been created!

"What remedy, if any, does the Nevada cattle rancher have? Legislation? Perhaps. Litigation? A class action suit might be brought in this court by the ranchers and by the State of Nevada on behalf of all wildlife against the Secretary and the BLM."

In the fall of 1973 the BLM notified ranchers throughout the West that they had a two-month period during which they could make a formal claim to remove privately owned horses and their offspring currently grazing on public land. A state brand inspector and representatives from the BLM and the U.S. Forest Service were to certify that the claims were legitimate. After November 15, 1973, all unclaimed horses would henceforth belong to the federal government and be protected. By the end of 1973 almost seventeen hundred people, most of them ranchers, filed ownership affidavits for more than seventeen thousand horses. Over seventy-five hundred were claimed in New Mexico alone; Nevada was not far behind with sixty-five hundred. Oregon, California, and Wyoming accounted for most of the rest.

From the ranchers' point of view, there were several advantages to claiming as many horses as possible: fewer horses meant more forage for cows and sheep; the more horses removed the less they would have to worry about their numbers increasing in the future; and, depending on roundup costs and market prices, there was money to be made in selling the claimed horses. Working in the ranchers' favor were state brand inspectors: many were ranchers themselves, and most were sympathetic toward the livestock industry. Militating against ranchers were fears that the BLM would "trespass" or charge them for not paying for the years that the animals ate and drank at taxpayer's expense. Trespass not only results in a fine, but also in a potentially devastating blot on a rancher's record. When four "willful trespasses" have been filed against a rancher, the BLM has the legal right to revoke his grazing permit.

In retrospect, it was clear that ranchers filed claims for horses based on their reading of two major factors: the likelihood of the BLM charging them "excessively" for past grazing by the horses, and whether, after trespass and roundup fees, they were likely to sell the horses at a profit. Apparently, few ranchers, and not many BLM employees, gave much thought to what would happen if horses were not removed. One exception occurred in the BLM Lakeview District of southern Oregon where the district manager saw the potential for long-run problems. He made a special effort to sit down with ranchers and give them inducements to make full use of the law. Irrespective of how old claimed horses were, he agreed to charge them a flat trespass fee of thirty dollars per horse, and nothing for colts. From 1974 to

1980, almost sixteen hundred wild horses were removed from the deserts to the north and east of Lakeview under the 1973 claim provision. This was a significant dent in Oregon's mustang population. But it was nothing compared to what happened in Nevada's Elko County.

The BLM district manager in Elko in 1973 realized that he could save himself and ranchers a lot of headaches and lawsuits if he could get all, or nearly all, of the county's wild horses removed. Unlike the Lakeview district manager, he had an additional incentive for rounding up as many horses as possible: a large share of the mustangs in Elko County lived on "checkerboard" lands. For roughly twenty miles on either side of Interstate 80, through virtually the entire state, just about every section or square mile is alternately public and private land. In the 1860s, the federal government granted five million acres to the Central Pacific Railroad in return for building the first transcontinental rail line and for free transport of government materials in the following years. Thereafter, three and one-half million acres were sold to ranchers and other individuals by the Central Pacific and its successor, the Southern Pacific. Much of the remaining land belonging to the Southern Pacific is leased for grazing. Since wild horses only have a right to roam on public land but obviously cannot distinguish between public and private domains—most of the checkerboard is not fenced—this presents the BLM with a thorny problem. Whenever the horses wander onto private property, ranchers can notify the federal marshal and demand that the horses be removed immediately.

The BLM manager urged ranchers in and around Elko to think hard when filing claims for horses. He declared that no horse on public land in Elko County was more than four years old, and that the most any rancher would have to pay to claim a horse was fifty dollars.

Elko's ranchers, by some accounts more progressive than those elsewhere in the state, understood the predicament. They adhered to the law by filing claims that provided information on the age and color of each horse claimed, and there were attempts to match affidavit descriptions with horses actually rounded up. But before long it became clear that accurately reconciling one's memory with half-forgotten or never-seen horses was not easy. It was, however, not difficult to decide which way to err. As one knowledgeable BLMer told me: "Some ranchers would get thirty or forty offspring from half a dozen horses that had been out there for the previous five or six years. You didn't have to be a mathematician to figure out what was going on."

From 1974 to 1978, more than thirty-five hundred horses were rounded up in Elko County. According to official BLM records, every last wild horse was removed from the county's checkerboard; when the last roundup was made, there weren't all that many to be found elsewhere in northeastern Nevada.

While all this was going on, detailed summaries on the number of horses rounded up, lists of who had claimed how many, and when each roundup was to take place were supplied to interested groups. The wary were even invited along on mustang gatherings. It was only after the fact that wild horse activists felt that they had also been invited to witness a hanging—and had been the ones hanged.

The T Quarter Circle Ranch, with headquarters along the Humboldt River three miles west of Winnemucca, is one of Nevada's larger checkerboard outfits. It sprawls for fifty miles through two counties, from Imlay on the west to Paradise Hill on the east, and for twenty-five miles north of Interstate 80. The ranch has a BLM permit to run just over two thousand cows on 225,000 acres of public land. With good summer pastures in mountains at either end of the allotment, and naturally irrigated land along the Humboldt River, cattle feed has not been a problem. The land also supports a healthy population of wild horses.

Hank Angus, whose family started the ranch in 1913, guesses that in 1971 there were no more than two hundred mustangs on the entire spread. Only once prior to 1971 did Hank try his hand at rounding them up. That year he caught a few and sold them in Reno, but in the end he didn't figure the effort had been worthwhile. Today he says, "I didn't get that many and those I got didn't go a thousand pounds. At four cents a pound, I lost money on the venture. Why bother?" After that experience he concluded that without proper traps, it was better to leave the job to outsiders. Periodically, freelance horse runners got permits from the Humboldt County Commissioner and then, with Angus's okay, came onto his ranch to take what they wanted.

Hank Angus says that in 1973 when the wild horse claim period was declared, the BLM in his district set the trespass fees so high "that there was no way you could afford to claim the horses. We heard the trespass fees were going to be $180 per horse, and we figured the horses were maybe worth $120 apiece." Then, as he sees it, to add insult to injustice the BLM district manager let it be known that by claiming the horses, the ranchers admitted that they had been running in trespass for years. In effect, the BLM was saying that ranchers were raising horses without paying the public for the use of the range.

And in many cases, the BLM was right. This charge rankled Angus and other neighboring ranchers who had never had much interest in the horses as a profit-making venture. The threat of willful trespasses meant that their livelihoods were in jeopardy. Angus's fears were deepened because he believed that the Winnemucca district manager was out to get him and other ranchers. "He'd do what he could to put us out of business," Hank believes. Despite the fear of reprisals, Angus went to the BLM state office in Reno on three different occasions to complain about the district manager and the problems he was having with horses. When he couldn't get a nickel's worth of commitment on anything out of the state BLM director, he finally said to himself, "The heck with it. Leave the horses out there."

In 1977, the T Quarter Circle Ranch received a willful trespass on its grazing record. That year the BLM had decided it was time to have all cows in the Winnemucca district ear-tagged in order to make it easier for the bureau to determine whether or not a rancher was grazing more cows than allowed by permit. In the same year that the T Quarter Circle was being required to ear-tag its cows, the Nevada BLM cited eighty ranchers for illegally running more than twenty thousand head of cattle, sheep, or horses on public lands. And this, BLMers believe, is only a small fraction of the cheating that goes on in a given year. Nevada stands first on the list of states in violation of grazing laws by ranchers.

Frosty Tipton, the young, vocal son-in-law who helps run the ranch, says that the bureau didn't give them enough tags for all of their cows. "Also, cows lose tags and it's hard to convince the BLM that this actually happens. The BLM says there's less than a one percent loss. That's not true. From our experience, about eight percent of the cows lose their tags." When government cowboys counted the T Quarter Circle cows, they identified a couple of hundred that had clean ears. "Then," Frosty explains, "when we made waves and complained that they had overcounted cows without tags, they dropped the number but gave us a willful trespass. Boy, that's the last thing we needed."

By this time the growth in horse numbers was being felt at the T Quarter Circle. Areas normally abundant in Indian ricegrass, Thurber's needlegrass, needle-and-thread grass, and Nevada bluegrass were beginning to look thin and bedraggled, and they were under invasion from weeds and less desirable annuals. By the middle of June the ranch's cows were no longer staying in their high mountain pastures as required by the BLM and the ranch's own rotation plans. From April through July, roughly half of the livestock was supposed to remain on the slopes of Eugene Mountain, near the western edge of the ranch. The rest were pastured far to the east on China Garden Mountain. The force behind the straying cattle was wild horses. Instead of a handful of them coming into the springs and wells for a drink and then leaving—as they had traditionally done when few in number—horses were now at water troughs for several hours at a time. The cows, no match for the horses and smart enough not to do anything too dumb, would wait long hours to get their share of water. When it finally became apparent that they would have to wait much of the day every day, and that they would get less than the ration they were accustomed to, they decided to leave. En masse, like a mighty army returning to base camp, they headed for home. When the Anguses woke up one morning, they saw that they had a yard full of cows. By BLM regulations, cows outside their bedroom window in the middle of summer is against the law.

Then there were the problems that the Fallinis knew all too well. When the horses couldn't get enough water, or get enough of it fast enough, they began smashing water troughs. Part of the horse behavior was, undoubtedly, playful—colts learning how to be macho studs, or how to fend off an unwanted sexual advance. Whatever, the playfulness also took place around gas pipelines. "Then before you knew it," Frosty says, "we had gas running and shooting out all over the place. We had to make all kinds of repairs." In addition, the horse population had grown so large that it required two daily visits rather than one to keep the well pumps going. For a while, Hank or Frosty would drive out to some of the wells early in the morning to scare the horses off so there would be a little water left when the cows arrived. This wasn't exactly legal, but the Anguses didn't feel that they had much choice. Still, their outlaw efforts didn't amount to much. The horses got wise and would just watch on a hill a couple of hundred yards away until Hank or Frosty left. Then they'd head down to the well, crowd out the cows, and tank up.

The Angus's complaints weren't heeded until 1981, when the BLM rounded up seven hundred fifty horses on the T Quarter Circle Ranch. Though there were notable changes, the drawoff wasn't enough: there were still more than one thousand mustangs left. Hank and Frosty continued to complain to the BLM, but to no avail. Or rather, nothing happened until they hired expensive lawyers and brought suit against the BLM. Then, suddenly—just as other ranchers have discovered—the Nevada BLM responded by placing the T Quarter Circle Ranch near the top of its priority list for wild horse roundups. The government gathered over one hundred mustangs in

1983, and it agreed to take off another five hundred by 1987, at the rate of about one hundred twenty-five a year.

Now that their wild horse problem is being solved, the Anguses feel better. Still, there are open wounds. Jane Angus, not one to mince words, speaks for Hank and Frosty. "You know," she says, "those BLM people don't hear you and it's real sad. We like horses, but why do we need this many? We're the only ones that deal with them and they just infuriate us. I might even see it having those horses out there if tourists would come by and see them. But why in the world do you want all these horses when no one, no one in the world can see them? Except for us and the BLM no one even knows they're there. And let me tell you something else. We love the game out there. Townspeople! We don't want them hunting here. We've got geese and deer and kit foxes and chukars on the ranch. We've got three pairs of geese and they've hatched thirty-one young so far!"

The wild horse experience has deepened the gulf of mis-understanding between the Anguses and the BLM. In Jane's mind, many of the problems they've had result from "high-falutin' eastern-ers" trying to tell western-bred ranchers how to run their businesses. "We had one BLM employee that came here from New Jersey. He had no idea what this country was like, and when he got here it scared him to death. He thought it was going to be like Kansas or something. We wouldn't have any idea about how to run their lives. You know, we get more and more people out here who have no idea which end of the cow the hay goes in. It's amazing, just amazing. And then some of them don't really understand what their job is. They want to manage your ranch instead of managing the range. I don't think people back east know anything about what we're doing out here. We're not here to rape this country. It's ours and we're not going to hurt it."

Younger members of the BLM in Winnemucca seem largely un-ruffled by criticism from the Anguses. Maybe it's their bureaucratic, public posture that I saw. Or maybe they believe that the Anguses have good cause to complain. Several of them have high praise for the T Quarter Circle Ranch. They see it as a progressive outfit: one that is computerizing data to determine exactly which cows were bred by which bulls; one that is developing its own feed lot to fatten cows in the fall before sending them to market; and one that is implementing a better rest-rotation system.

Brent Eldridge is a rancher in eastern Nevada's White Pine County. When I talked with him in the summer of 1982, he was one of four directors of the Ely Cattleman's Association, and chairman of the Ely

Board of County Commissioners. He was pressed for time; meeting in his noisy commissioner's office, he spread his thick arms, leaned toward me, and said, "Let me begin by saying it would be hard to find a single rancher who wouldn't say he'd like to see a few horses out there. But I will guarantee you every last one of them will say there are too many there now. Way too many."

He didn't see much evidence that the BLM was managing the wild horses. "Which is their job, presumably," he said. "Because they're not doing anything, the condition of the range and the condition of the horses is going to hell. The problem is those horses are out there twelve months a year, and no matter how the sheepmen and cattlemen are managing the range, the horses are defeating them."

As Brent Eldridge got up to run to an appointment, he said, "I personally know a Pan Am pilot in Portland who flies to France a couple times a week with loads of frozen horsemeat. Hell, lots of people eat horsemeat. If you get it young enough or old enough and cook it right, it's as good as beef. And better for you too." There was a brief pause, then he said, "The government wouldn't have horses to protect if the livestock people didn't want to see them on the range. *We* put them out there and *we*'ve kept them there, but never in unrealistic numbers."

It was said that in Rock Springs, Wyoming, at the end of the nineteenth century you could stand on a high hill in early October and as far as the eye could see there would be sheep camps. Most of the sheep belonged to "foreigners," sheepherders who drove their bands in from Idaho, Utah, and far-western Wyoming. At that time, perhaps eight hundred thousand to a million sheep moved through this area each year. Then suddenly, the endless white waves of wool on the hoof coming in early to take prime grasses were more than Rock Springs sheepmen could tolerate.

In 1901, the Sweetwater County Range Association was formed, and its first order of business was to lease from the Union Pacific Railroad its odd-numbered sections in the checkerboard, a seventy-by-forty-mile swath that extends from the town of Green River in the west to Tipton in the east. The association got what it wanted: a million acres at a penny an acre a year. Combined with the even-numbered sections, another million acres that belonged to the state and the federal government, the association felt that it had effective control of the checkerboard. And indeed it did. The first out-of-state sheepherders who tried to avail themselves of the rich grasses along the Green River and Bitter Creek in Sweetwater County were thrown in jail and their sheep were impounded. Soon thereafter, the Supreme Court upheld the right of the association to prevent anyone from crossing private land (leased from the Union Pacific) to get to public land. Then in 1907, the Sweetwater County Range Association further consolidated its control of the area when it formed a sister organization, the Rock Springs Grazing Association. The RSGA promptly leased or bought outright more land from the railroad, and sold shares to its members, which gave them the right to graze in proportion to how much stock they held.

One of the founding members of the RSGA was John Hay, who for decades had one of the largest sheep operations in all of western Wyoming. At one time he had upward of thirty-five thousand head. Like other sheepmen, Hay required lots of horses, forty to fifty for a single sheep camp. Each camp had between two thousand and twenty-five hundred sheep. Hay bred and raised most of his horses. For much of the year they grazed on the open range, and then in the fall or spring, he rounded up as many horses as he could, branded them, broke and gelded some of the stallions, and released the mares. Because he and other sheepherders needed large horses for packing salt and supplies and for hauling wagons, they upgraded their free-roaming horses by periodically putting out high-quality Belgians and Percherons. If a rancher spotted a small or bad-looking stallion that couldn't be caught, he shot it.

When World War II came along, sheep ranchers couldn't get enough help to run their operations. Even substantially reducing the number of sheep didn't solve the problem for many and, as a result, wranglers who had previously spent much of their time rounding up and breaking horses now had to be used almost exclusively on the sheep stations. Soon, the horses began to multiply; they became increasingly wild and difficult to catch. The labor problem didn't improve even after the war, and to keep the herds from becoming excessively large, ranchers hired fixed-wing pilots to round up the wild horses. Attractive stallions were still gelded for ranch use, but more were made available for sale to rodeos and pet food factories.

When the 1971 Wild Horse and Burro Act was passed, its meaning was all but lost on most members of the Rock Springs Grazing Association. They simply assumed that the BLM would control horse numbers just as ranchers had done for decades. But in these early years the Wyoming BLM, like the BLM in Nevada and Oregon, had no clear vision of how to manage wild horses. By 1976, it was evident that Wyoming ranchers had been too optimistic. Horses seemed to be everywhere, especially around water, and especially in the checkerboard during the winter when they moved south for protection and more accessible feed. Whereas some sheepmen had been running as many as fifteen thousand sheep through sections of the southwestern Wyoming checkerboard prior to 1971, they could only take eight thousand to ten thousand through the same areas a decade later. The reason: too many horses eating too much grass.

The ranchers contacted the federal marshall. They demanded that the horses, which by law they no longer owned, be removed from private land—those lands owned or leased from the Union Pacific. It was, they rightly complained, utterly impossible to keep the mustangs on the public sections; probably no area in the West has fewer fences. Months rolled on and nothing happened. The federal marshall was again sent written demands that the horses be removed, and still nothing happened. Nor did the RSGA have any luck with Neil Mork, then BLM district manager for this part of Wyoming. Contrary to a long history of dealings with the Grazing Service and the BLM, the ranchers finally met a BLM boss who didn't wear sheepskin. Mork was, by many accounts, brash, even antagonistic toward the ranchers. And when it came to wild horses, Mork had definite ideas on how to manage them.

By 1977, the RSGA ranchers decided that the only recourse they had to prevent mustangs from eating a sizeable share of grasses they paid for was to sue the BLM. They hired James Watts's Colorado legal firm, Mountain States Legal Foundation, to handle the case. Under Watts's direction, the suit against the BLM contended that "wild horses had caused deterioration of forage, water, topsoil and property values on private lands," and it was therefore the federal government's duty to thin wild horse herds in southwestern Wyoming down to their 1971 numbers. The ranchers also asked for damages for all grasses eaten by the horses. They claimed that there were more than six thousand mustangs giving them trouble, in their estimate more than a three-fold increase in the population since enactment of the 1971 law.

At about the time the lawsuit was filed, four women in Green River, twenty miles west of Rock Springs, heard a rumor that the BLM had plans to remove all mustangs from the checkerboard. They found it objectionable, not only because they considered the checkerboard horses to be among the most accessible in the state for public viewing—Interstate 80 cuts directly through the middle of the checkerboard—but also because they had heard another rumor, that Neil Mork wanted to create a preserve for up to six hundred wild horses in an isolated badlands south of the interstate near the Colorado border, in an area called Adobe Town. Since Adobe Town is really only open to four-wheel drive, only a small fraction of people with an interest in seeing mustangs would be able to do so. Mork believed that his solution would take care of the checkerboard problem and also make it easy for the BLM to manage the horses.

The Green River women, looking for the most effective way to combat the BLM, joined forces with John Borzea, a Rock Springs resident who had been interested in conservation and wild horse issues since the early 1950s. In 1955, Borzea, a gas plant operator and part-time real estate dealer, almost single-handedly got the BLM to put a couple of hundred elk in the Red Desert north of Rock Springs. For many years he was secretary, then president, of the Sweetwater County Wildlife Association, and in 1964 was named Wyoming's "Conservationist of the Year." Borzea's concern for mustangs began in the early 1950s when he became interested in how they were being rounded up. From time to time, he'd go down to the Union Pacific Railroad yards and see crippled and maimed mustangs and, occasionally, drunk cowboys trying to ride them. Borzea learned that colts were left behind by pilots because they were slow and not meaty enough, and he discovered that lead mares or stallions were occasionally gunned down if a pilot was having trouble turning a herd. For years, he tried to persuade ranchers not to use airplanes for their roundups, but he had no success. The real turning point in Borzea's involvement with wild horses occurred in the mid-1950s when he came upon a makeshift mustang slaughterhouse five miles west of Rock Springs. "*That*," he says, "is what really got me involved. That place was only in operation about a year, but from then on, I wanted to keep a much closer eye on everything done with those horses."

By this time, others were beginning to show concern for the mustangs. In 1957, the Wyoming legislature removed the mustangers' profit incentive by passing a state estray law declaring that unclaimed, unbranded horses had to be sold at public auction. If a rancher wanted to round up mustangs, he had to bear his own roundup costs, transport them to the auction house, pay for their feed, and then buy them on the open market. To no one's great surprise, ranchers lost interest in horse roundups, and the wild horse problem worsened.

Before long, the Green River women, John Borzea, his wife, and a couple of other Rock Springs locals had formed WHY—Wild Horses Yes. Cognizant of the poor image of the BLM, the fact that many townspeople were strongly pro-wild horse, and that key members of the RSGA wanted to maintain a good image in this part of the state, WHY and the RSGA agreed to meet to see if something could be worked out. Neither WHY nor the RSGA was interested in BLM participation. The RSGA was particularly hostile to the bureau, feeling, as one member put it, that "The BLM is a typical government bureaucracy. It's very seldom in a position to sit down and say that it agrees to anything. We figured that bringing the BLM in would have made it impossible to get any kind of an agreement. We had a problem, we couldn't wait forever."

Among those present at the joint meetings of WHY and the RSGA was John Hay's son, John Hay Jr., who at this time effectively ran what was left of his father's empire. Though willing to compromise, he made it clear that federal law notwithstanding, all of the horses on the open range technically belonged to the ranchers. If anyone doubted his claim, he'd provide a list of who owned what and exactly how many horses each and every rancher had been running on the range during the previous half-century. John's brother, Leonard, was vocal on another matter. He said that he and other ranchers saw horses as a commodity, no different than cows and sheep. They had to be harvested, and there was absolutely nothing wrong with sending them to slaughterhouses where they could be sent to Europeans and others who eat horsemeat. Other RSGA ranchers expressed concern about their image. They felt that too many people, especially those who had recently swelled the population of Rock Springs because of oil and gas discoveries, were ignorant of the history of ranching in the area. The newcomers didn't understand the damage done by horses, and they didn't appreciate that ranchers were "good stewards of the land" who were intimately familiar with the desert's carrying capacity and therefore not about to overgraze. But, just as the members of WHY didn't want a zoo in Adobe Town, so the RSGA didn't want one in the checkerboard. The ranchers (at least the majority of them) merely wanted to be "reasonable" about the horse problem. They wanted to reduce the number on the checkerboard and other lands that horses had traditionally grazed to what it was in 1971, which by their calculation, was around twelve hundred.

Working in WHY's favor was the fact that the tiny organization was widely perceived to represent public opinion. Borzea and a couple of others also knew that while a few members of the RSGA would be happy to see every last horse removed from the checkerboard now that they no longer owned them, most ranchers were reasonable and would probably go along with a reasoned solution. WHY proposed that a total of eighteen hundred wild horses be left dispersed over an area that the BLM estimated had more than two thousand in 1971. The RSGA wanted no more than twelve hundred horses in the area, including a maximum of five hundred on the checkerboard. By their second meeting in January of 1979, WHY and the RSGA agreed to split the difference. They agreed on a total of fifteen hundred mustangs for an area that covered the RSGA checkerboard plus large hunks of desert to the north and south. Of this total, only five hundred would remain on the checkerboard—three hundred north of Interstate 80 and two hundred to the south. One additional proviso was acceptable to both parties: if more than five hundred horses were counted on the checkerboard in summer, the excess would have to be removed. The ranchers didn't care how many mustangs used these lands in winter, even though winter is the time of year when their sheep make exclusive use of the checkerboard. The ranchers knew that in most years, the hard winters on the open deserts north and south of the checkerboard force mustangs into the shelter of the canyonlike lands along Bitter Creek.

For several years, the Rock Springs BLM office had been working on an environmental impact statement and a management resource plan to determine the ideal number of cows, horses, and wildlife for this part of the state. Other than Mork's idea of clearing horses off the checkboard and putting many of them in Adobe Town, apparently no one in the BLM had much of an idea on how to allocate feed among the various competitors. It is not surprising, then, that when the BLM was presented with the agreement between WHY and the RSGA, it readily agreed that fifteen hundred horses would fit in very nicely with its management resource plan. It also happened that when a federal court judge handed down a decision against the BLM in the checkerboard case, the agreement between WHY and the RSGA became the basis for determining the "proper" number of horses to be left on public lands used by the RSGA: fifteen hundred. The judge gave the BLM two years to remove all horses above that number.

Although the RSGA would have liked the horses rounded up at a much faster rate than ordered by the court, generally, most of its members were pleased with the outcome. A few could even smile wryly when, in 1981, one RSGA range rider saw the BLM returning twenty-five mustangs to checkerboard lands. It seems that in its eagerness to comply with the letter of the law, the bureau had taken off a few too many horses. Not wanting to be embarrassed, the BLM district manager ordered the horses returned. It apparently didn't occur to anyone that the horses would have naturally recouped their number in the next foaling season.

In a sense, the agreement between WHY and the RSGA was a victory for everyone concerned. It absolved the BLM of the hard work of justifying how many horses would be proper for this part of Wyoming. It gave large numbers of locals confidence that the ranchers were not ruthless plunderers who cared not a whit for mustangs. It got the wild horse population down to a figure that most members of the RSGA thought they could live with. And it satisfied those in WHY who had fought for the horses and who, in John Borzea's words, feared that "without an agreement the ranchers were in a very good position to get total removal of horses from the checkerboard."

Shortly after the agreement between WHY and the RSGA was signed, WHY disbanded. Borzea and others felt that its mission had been accomplished. Still, there are those in Rock Springs and Green River who continue to keep an eye on the BLM. They say that they're not certain that the bureau always knows the difference between a solution that momentarily satisfies and one that is best for wild horses.

CHAPTER 5

VOCAL CRITICS

In the spring of 1977, the BLM counted almost nine hundred mustangs in the Buffalo Hills, a game-rich hunk of ground between Poodle Mountain and the California border north of Reno. The horses shared the hills with mule deer, coyotes, several species of rodent, one of the largest wintering populations of antelope in the state, and more than a thousand cows. In all, an excess of mouths for the lean desert forage. As one BLM range specialist familiar with the area put it, "There was simply no way there were enough groceries for everything out there." What that same BLMer did not say was that all the signs of impending disaster were there—and they were not heeded.

It took the winter and spring of 1978 to convince the government that they had more than a potential catastrophe on its hands. At least three hundred wild horses starved to death; no one knows the real number. Rodger Bryan, a BLM range specialist familiar with the Buffalo Hills, told me: "We're lucky if we counted fifty to sixty percent of what actually died." Bryan is firm in his opinion that the area had shown definite signs of overgrazing for some time. "If the hills had had half the population of horses and cows that it did, the dieoff wouldn't have occurred. We should have taken some of the horses off."

"Why didn't you?" I asked.

"Because some wild horse advocates wouldn't allow it." He said that most activists are suspicious of all attempts by the BLM to reduce the number of mustangs. In the case of the Buffalo Hills, they demanded—and didn't get—detailed documentation of resource damage.

As in many animal populations, the young and the old were the first to die. Almost sixty percent were under four; another twenty-five percent were ten or older. Had it been a severe winter, with biting cold and brisket-deep snow, one can only guess how many wild horses might have perished. After the disaster was recognized, the BLM did go into the hills with a helicopter and cowboy crews and reduce the remaining horse population by several hundred. The BLM also demanded that the number of cows be reduced; it then eventually closed the hills to all livestock grazing for two years—more than three years after the fact. Even at the time of the tragedy, however, the BLM effort was too late for a long list of living things.

Lots of mule deer and more than one hundred cows died in the Buffalo Hills that fateful winter. The cows, the BLM maintains, didn't die from lack of water, but because their owner had not been adequately feeding them. At one point, after neighboring ranchers described some of the cows as "walking bags of bones," the bureau took it upon itself to impound four hundred of the underfed animals. Not content with the perennial use of Indian ricegrass, Idahoe fescue, needle-and-thread grass, rye grass, and white sage for the monthly per-cow equivalent of parking meter change, the rancher also from time to time grazed more cows on his public allotments than the fragile desert could withstand, or law permitted. In 1956, his grazing permit in the Carson City BLM district was suspended, and then revoked four years later because of repeated violations. He then moved part of his ranching operations into California's Susanville district. Once again grazing privileges were revoked for abuse of public land. While being prosecuted by the Susanville BLM, he found a lenient manager in the Winnemucca BLM office, which looks after the Buffalo Hills, to issue him new grazing rights. Before long, Nevada BLMers began complaining that they couldn't get him to put ear tags on his cattle so that the agency could determine whether he was grazing more livestock than allowed. Then in 1978, the year of the massive dieoff in the Buffalo Hills, an administrative law judge in the Interior Department found him guilty of further trespass violations in the Susanville area. This time he was fined several thousand dollars and his remaining grazing rights for that area were revoked. Why the Nevada BLM didn't cancel all of the rancher's grazing rights long before the winter of 1978 is a question for which there cannot be a flattering answer.

Nevada, like other western states, has had its share of ranchers who regard grazing permits as a right rather than a privilege. They hark back to a time prior to the 1934 Taylor Grazing Act when no one told them that they had to think about a range's carrying capacity. In many cases grazing allotments have been passed on through several generations, more than enough time to make one feel that continuous use confers private ownership, the right to unbridled use. What can the BLM do about it? It can, if it wants, take away a rancher's grazing permit and thereby end his cowboy career. But the BLM has been so constantly vilified and browbeaten that it seems to like itself best when hesitant and pokey, feigning ignorance of the obvious. The simple truth is that the bureau is notorious for letting troublesome, vocal ranchers have their way.

It would be easy to throw all the blame for the Buffalo Hills disaster on the callous rancher, or on the Nevada BLM for not putting him out of the ranching business a long time ago, or on the wild horse advocates who required documentation of resource decline before they would consent to a reduction in mustang numbers. But, as is so often the case, a complex web of factors contributed to the tragedy. First, there was a fence on the extreme western boundary of the Buffalo Hills, a barrier of sufficient length and strength to prevent the movement of one-third to one-half of all the hills' wild horses. Without the fence, many of them might have made it to lower ground; they might have found enough food—and left enough for others who remained behind—to make it through the winter.

Add in a bit more complexity. In the year of the mass starvation, the BLM had its many eyes telescopically focused elsewhere. The agency was facing another potential mishap, this one in the Owyhee Desert far to the northeast along the Oregon and Idaho borders. There, about eighteen hundred mustangs lived on 615,000 isolated acres. When a helicopter census was taken, the horses were spread throughout a good portion of the Little Owyhee's Spring Range. Two months later, forty percent of them had moved to Twin Valley Spring, a usually dependable watering hole. It was not dependable this year. It had begun to dry out, and when BLM cowboys arrived they found eight dead colts and a yearling buried in mud as thick as pea soup. In their eagerness to get to water, thirsty mothers and studs had pushed the colts into the stifle-high muck. The little ones couldn't get out, and though the BLM rescued a few of them, many promptly returned.

Precipitation that year was seventy percent below normal. Springs with water were controlled by ranchers and the horses couldn't get to them. Here as well, an eighteen-mile-long fence had been built at a ninety-degree angle to fifteen horse trails. The barbed fence had three gates; two were closed. All of them were supposed to be open throughout the winter and spring. At one spot, the horses tore down the fence for a distance of one hundred seventy-five feet because it was the only way to get to water. Elsewhere along the lengthy wire wall, one could find bent posts and loose strands and a wide variety of colors of horse hair on the barbs. Ron Hall, the biologist who made an assessment of the Owyhee problem at the time, declared that the fence was "an ecological disaster." He said, "If it was designed with the objective of having maximum negative impact on wild horses, it would be difficult to do a better job." Drought and a fence were only two of the problems. By BLM estimates, the Owyhee Desert did not have nearly enough forage for the mustangs, to say nothing of what was needed to feed rabbits, ravens, badgers, coyotes, field mice, and the estimated three hundred pronghorn antelope and one hundred-fifty deer that call this sky-filled desert home. In normal years, several thousand head of cattle graze the Owyhee. Luckily, they had been pulled off because of the severity of the drought. When the BLM finally realized that something had to be done, they made plans to round up eleven hundred horses.

Meanwhile, investigation of carcasses and dying horses in the Buffalo Hills revealed the extent of the tragedy. Piles of jet-black fecal material around the emaciated and chewed bodies indicated that the horses had been on their shoulders and flanks for as much as five days before they died. Some were lucky; their lives had been shortened by armies of voracious ticks. Summer and winter alike, horses occasionally practice coprophagy; that winter, feces may have been their main meal. In struggling to die, the Buffalo Hills mustangs ate manes, tails, patches of facial hair. Sisters, brothers, mothers were fair game. Some had backed themselves up against a bush or rock to protect hind quarters from the greedy mouths of coyotes. Mothers and fathers could only watch as their foals were a first meal, a mirror of their own fates. Some dug earthen ramparts, pitiable sodden cloaks that suggested they were aware of their lot.

The ice cream plants—the most desirable grasses and forbes— were the first victims of the catastrophe. In one area of the Buffalo Hills, Tin Canyon, the scanty vegetation was completely picked over, what the BLM refers to as "one hundred percent utilization." A dozen rabbits would have had a hard time eking out an existence. Six-foot-tall sage was stripped of bark from ground level to five feet, as though a blind carpenter had gone berserk with an electric sander. Tetradymia plants, or horsebrush, were kicked and pawed as horses fought for the coarse twigs and bark. Throughout the hills, grasses pedestalled; they looked as though a child had dug around their bases with a blunt knife and then—the roots exposed—mysteriously stopped short of finishing the innocent murderer's job. Contrary to popular belief and the misinformed premise of some environmentalists, wild horses, like other animals, have no social mechanism for behaving kindly toward their living food sources. Horses do not self-limit their numbers before inflicting significant and often permanent damage to their environment.

No autopsies were performed on the mustangs, so one cannot know how many died from locoweed and death cama. In normal years, these lethal plants do not pose a problem. But in a year like 1978, the animals eat just about everything. It only takes three or four pounds of death cama to kill an adult horse. The striking locoweed, with its dense spikes of violet flowers and fruit that resembles garden peas, can be even more dangerous. It produces frenzied behavior followed by partial paralysis, impaired vision, then death. Like mercury, it accumulates in the body, and kills a horse in less than a month.

Amazingly, by late 1983, there were perceptible signs that some of the Buffalo Hills grasses were recovering. The lucky ones, mostly annuals, were those buried among rocks or under heavy cover of shrubs. By 1984, the wild horse population was above three hundred and increasing. Soon cows would once again be allowed into the Buffalo Hills.

In the spring of 1983, I went to Washington, D.C., to talk with Joan Blue, president of the American Horse Protection Association. The AHPA, known to many in the Midwest and the East for its active role in looking out for the welfare of Tennessee Walking Horses and exposing the use of drugs in race track horses, has been the nation's most vocal critic of the BLM and its handling of wild horses. Among western ranchers, mention of the AHPA provokes cursing, unflattering comments about meddling easterners, and vague references to the Georgetown resident whom just about everyone refers to as Mrs. Blue.

After Joan Blue and I talked for an hour or so, and after she had left me with the impression that for her no number of wild horses would be too many for western rangelands, I mentioned the Buffalo Hills incident. She said that she had seen photographs of the calamity, then added, "I don't believe it happened. No one ever came up with anything but the pictures. They had dates and places on them, but those could have been horses that died any place at anytime. I'm well aware of the treachery of BLM propaganda. They have lied to Congress, and they've lied to the public. So what would a couple of pictures be if that would help them along?"

I said that I had qualms with some BLM practices, but that after talking to scores of range personnel and district managers and those charged with caring for wild horses, no more than one or two of them had struck me as purposely crooked of mind. I said that I had talked with several people who had firsthand knowledge of the Buffalo Hills incident, and though there were minor disagreements about some of the facts, there was no doubt that the dieoff had occurred. Joan Blue dismissed my objections, then followed with a question about my real motives for coming to see her.

Joan Blue, whose training is in political science, is unalterably convinced that (in her words) "The BLM is really out to get the horses." She believes that the bureau's overriding concern is to put them all on a commercial auction block and send every last one of them to a slaughterhouse. This preoccupation, based on what she describes as "big business greed and big government bureaucracy," is often vented in AHPA Newsletter headlines: "The Wild Horse: An 'Endangered Species'"; "Wild Horse Slaughter Law Threatened"; "Adopt-A-Horse: One Way Ticket to Slaughter."

"How many wild horses do you really think the BLM would like to see in the West?" I asked.

"None."

"Literally?"

"Yes, literally. Those who work for the BLM are just a bunch of grass growers, graduates of Brand X College. They're not interested in animals."

I tried to talk with her about a number of wild horse issues, but Joan Blue seemed primarily interested in manifesting her contempt for the BLM and for ranchers. She wanted to let me know that the AHPA, with its fifteen thousand members, is "an outraged bunch, who enjoy nothing more than writing to their senators and congressmen to let them know what they think."

Joan Blue is absolutely certain that the nation's wild horses have never been counted "scientifically." What she means by scientific is that the horses should be tallied on foot. I was unsuccessful in putting forth the argument that counting horses on foot is not only difficult but invariably results in undercounting. Nor was I persuasive in arguing for the inadequacies of doing census work on horseback or in a four-wheel drive truck. She is certain that using airplanes and helicopters—easily the most reliable way of doing the job—produces results that are more than suspect. She believes that the BLM consistently exaggerates the number of wild horses running free in the West.

I mentioned a National Research Council report prepared by more than a dozen professional biologists, in which it is stated that the BLM's figures on mustang numbers are conservative: "There are more wild horses on western public lands than is generally assumed to be the case . . . and hence more range forage is being consumed by them than is generally believed." Perhaps the single most striking example of the BLM's conservatism is found in the bureau's often-used and commonly accepted figure of seventeen thousand wild horses for the entire West in 1971. It now looks as though this number is too low by five to ten thousand. Perhaps more. The proof is in increasingly careful censuses conducted by the BLM, in studies done by the University of Minnesota showing systematic undercounting by the BLM, and in actual numbers of horses rounded up and adopted out—figures which have led all but a small minority within the BLM to conclude that horse herds have been multiplying at between fifteen and twenty-five percent a year. These are population growth rates which, for horses, are all but biologically impossible, even in the most favorable environments. Poor census methods in the early 1970s are virtually the only way to account for the startling number of wild horses that the BLM and scientists can now document.

As far as I could tell, Joan Blue has been of one mind on the habits of the BLM and western ranchers since she became president of the AHPA in 1975. Shortly after taking office, she wrote to her membership that "until AHPA's legal and legislative program produces positive results" to make wild horse protection laws stronger, the AHPA "intends to take the Bureau of Land Management to court over each and every proposed roundup." Time would show that Joan Blue has been true to her words and that she's not one to waste time. In 1976, the AHPA and the Humane Society of the United States brought an injunction against the Idaho BLM to halt the state's first government gathering of wild horses since passage of the 1971 Wild Horse and Burro Act. The injunction and subsequent lawsuits would prove to be one of the AHPA's most notable, most visible, and most prolonged fights with the BLM.

In the mountains of eastern Idaho along the east fork of the Salmon River, not far from the tiny town of Challis, is a 260,000 acre, irregular rectangle of high desert that the BLM refers to as the Challis Planning Unit. It is a rolling, rounded, here-and-there jagged landscape patched with Douglas fir, mountain mahogany, lots of sagebrush, and a variety of grasses and forbes that feeds sizeable populations of bighorn sheep, antelope, elk, deer, chukar, bobcats, mourning doves, golden eagles, several kinds of grouse, a couple of bears and mountain lions—and cattle and wild horses. In 1971, a BLM aerial census counted one hundred-fifty mustangs in the Challis region. By 1976, another tally showed that the figure had jumped to four hundred ninety. Based on this and other interim surveys, there were strong indications that the 1971 census considerably underestimated the number of horses in the region.

Because of a successful suit brought against the BLM by the National Resources Defense Council in 1973, which claimed that the government was gathering insufficient information on livestock grazing and its effects on the public domain, the bureau was ordered to prepare two hundred twelve environmental impact statements for western rangelands. The very first of these was conducted on the Challis area, and among its many findings was that the range could only support three hundred horses on a year-round basis. Summer feed was sufficient for five hundred seventy-five, but in winter there was adequate forage for little more than half this number, because about half of the area is covered in deep snow. Furthermore, deer, elk, and antelope need the range's grasses and forbes to make it through the winter, and this is the time of the year when the horse diet is highest in browse-type plants. The environmental impact statement also noted that when grazing is too intense, plants don't get a chance to recover. Once an area is overgrazed, soil erosion occurs more rapidly, the water table drops, and then sage grouse and other small animals that depend on succulent vegetation disappear. In the area used by horses at this time, less than ten percent of the land had a stable watershed. Almost sixty percent had already suffered moderate to critical damage. Because of their large hooves, horses have the undesirable effect of increasing soil compaction, especially in wet areas. With these facts and ecological principles in mind, the BLM concluded that it should hold wild horse numbers somewhere between a minimum of one hundred eighty-five and a maximum of three hundred. This meant that more than two hundred horses would have to be gathered and put up for adoption.

A couple of weeks before the roundup, with the traps in place, a helicopter pilot contracted, and the cowboys hired and ready to run horses, Gail Snider and Marilyn Hall, both representatives of the AHPA, went to the BLM in Salmon, Idaho, and said that they had seen a mere twenty-two wild horses in the Challis area. They seriously doubted that there were four hundred ninety as claimed by the bureau. Sensing trouble, the BLM—at taxpayer's expense—took Marilyn Hall up in a helicopter. In an unsystematic "tourist" flight that lasted about an hour (a normal census for this area would take three to four hours) and covered about half of the wild horse area, Marilyn Hall counted two hundred thirty-two horses. The BLM wild horse specialist who sat beside her saw two hundred seventy-one.

The day before the roundup was to take place, the AHPA and the Humane Society of the United States got a federal judge in Washington, D.C., to stop the roundup on the grounds that it was "arbitrary and capricious." The judge said that the gathering would be inhumane, as the nearest veterinarian would be thirty miles away from the roundup site; the BLM didn't have enough credible information on herd growth rates; and the bureau's environmental impact statement was incomplete. In the injunction, the AHPA and the HSUS contended that there were only two hundred seventy-one horses in the Challis area; the AHPA felt that the area could support many more. Meanwhile, the HSUS had circulated fliers nationwide asking people to write to the Secretary of the Interior in order to "stop the indescribably cruel and unnecessary roundup of two hundred sixty wild horses in Challis." The flier went on to say that, "Incredible as it seems, after Interior removes half of the wild horses on that range, alleging that they eat too much grass, they intend to add more cattle to the four thousand already grazing. It is a windfall for cattle owners."

Among those who testified for the AHPA and the HSUS was Lorne Greene of "Bonanza" and Alpo dog food fame, later to become vice chairman of the Board of Directors of the AHPA. Greene said that the horses had to be protected, that they were being "sacrificed" for the four thousand cows that periodically used the range. Greene and the court were unmoved by testimony from BLM range specialists and the Idaho Director of the BLM who pointed out that the Challis cattle only graze the range from one to four months a year, while the horses are out there all year round. Nor did it seem to affect the court when it was pointed out that wildlife and domestic livestock numbers had been held constant since 1971, managed by hunting licenses and grazing permits, while the number of horses had increased dramatically.

The plaintiffs argued that if preservation of the winter range was at issue, then the BLM should have fenced it off. The bureau replied that it would require seventy miles of fencing at a cost of four hundred thousand dollars, which was prohibitive. In addition, fencing would block the seasonal movements of wildlife and horses. The court sided with the AHPA and the HSUS, ruling that the BLM did not have a good handle on counting horses, that the Challis region could support more of them than the BLM contended, and that fencing of the winter range should have been given more careful consideration by the government.

The wild horse population continued to grow, and the BLM continued to push for roundups. New injunctions and new lawsuits were filed against the BLM. New issues were introduced: the origin of the horses, how much they weighed, how much they ate.

In some of these court battles, the AHPA and the HSUS had another star witness, Hope Ryden. A preservationist and New York journalist who yearly goes to Montana's Pryor Mountains to photograph the reserve's one hundred fifty or so wild horses, Hope Ryden frequently testifies as an expert witness on behalf of the AHPA. She asserted that the Challis horses couldn't possibly be eating as much as the BLM claimed. The reason: they only weigh eight hundred pounds on average, she said, rather than one thousand pounds, the figure put forward by the BLM. Ryden's argument rested, in part, on an appeal to "climate and geography." She said that the cold, northern clime would work against nature selecting for a one thousand pound animal. (Apparently Hope Ryden was unfamiliar with Bergmann's Rule, which states that in warm-blooded animals body size increases with a decrease in average temperature. Large size in a cold climate has adaptive value in preventing loss of body heat. This explains why polar bears weigh up to eleven hundred pounds and Alaskan brown bears even more, and why deer are generally larger in the northern parts of their range than farther south.) In her court appearance, Ryden also said that she had been in the Challis region and could not find skeletal remains demonstrating that the horses were descended from draft animals. She said that the "small" horses were descendents of Spanish Barbs. As evidence, never seen by the BLM, she said that she had found an example of fusion of the fifth and sixth vertebrae. Fusion of these vertebrae, a phenomenon known as spondylitis, is a characteristic common to the small Spanish Barb.

The BLM countered that the Challis horses were the offspring of horses put out by ranchers as early as the late 1870s. By 1900, one individual was running three hundred to four hundred head on the open range. Many of them were never rounded up. Other Challis ranchers raised work horses, and they upgraded the quality of those running free by putting out good draft blood—Shires, Belgians, and Percherons. (Lingering evidence of Percheron genes could be seen in the 1970s in the size of the animals and in the fact that about twenty percent of the horses were grey.) In the 1920s and '30s, there were yearly wild horse roundups in the Challis region. Thousands were gathered and used either as saddle and work horses or trailed to Blackfoot and other railroad points for sale to the highest bidder. In 1952, the BLM contracted with private individuals to remove almost eight hundred mustangs. From then until 1971 when the wild horse act was passed, mustangers periodically came in to gather horses.

No one in the Idaho BLM had seen a wild horse that suggested Spanish origin. In court, the BLM apparently made no attempt to argue that the fusion of certain vertebrae may prove little more than that a horse has received prolonged stress.

Not content with the BLM's historical reconstruction and other evidence, the court ordered the bureau to weigh the Challis wild horses. It did so, and it found that the average weight of stallions was nine hundred ninety-three pounds, while that of mares was nine hundred ten pounds. These were figures gathered after capture and sorting, or after the horses had lost three to five percent of their normal body weight, which would put the average weight for adults around one thousand pounds.

The AHPA got other expert witnesses to testify that a one thousand pound horse would only eat seventy-three percent of that consumed by a nine hundred pound cow with calf. According to these witnesses, the BLM was overstating the food requirements of wild horses by around fifty percent. The BLM, then as now, claims that a horse eats about twenty-five percent more than a cow. The AHPA also had one witness state that a nine hundred pound cow was "unusual," that a western range cow weighs more like one thousand to eleven hundred pounds. The defense pointed out that cows on Idaho's rangelands are typically eight hundred to nine hundred pounds. Others noted that those who argue for higher figures have "a county fair mentality." Somewhere in the middle of the accusatory testimony, an "expert" witness for the AHPA also contended that a cow has much greater food requirements than a horse: the proof, this expert said, was in the number of teats! (Cows have four, horses two.) The real clincher for the BLM, however, came from evidence presented by three university nutritionists, all of whom had worked with mustangs. They testified that a wild horse of the sort found on the Challis range was eating about one thousand pounds of forage a month as compared with eight hundred for a nine hundred pound cow with calf. The expert witnesses for the AHPA had it backwards.

By the time the BLM was allowed to undertake its first roundup in the Challis region in July 1979, the wild horse population had passed seven hundred. That year, one hundred sixty-four horses were re-moved, and then another three hundred seven the following year. A study done in the area by Richard Hansen of Colorado State University showed that there was similarity in the diets of elk, wild horses, and cows. For cows and horses, dietary overlap ranged from sixty-one percent in the fall to sixty-eight percent in the spring. In good part to reduce pressure on the delicate Challis ecosystem, in 1979 the BLM also asked thirteen ranchers to reduce their cow herds, a few by more than fifty percent. The ranchers agreed on the condition that horse numbers would be reduced further, down to two hundred.

The Challis wild horse issue remained a sore point with the AHPA. The protectionist organization continued to insist that the BLM was overcounting the horses, that the carrying capacity of the range was much greater than the BLM maintained, and that the government was not doing nearly enough to solve the problem. The BLM, as it happened, was preoccupied with other problems; one in particular now stands out in the minds of Idaho BLMers. In the fall of 1980, as part of a range improvement program, the bureau burned several thousand acres of sagebrush in the Challis region. This, in theory, would allow forbes and grasses—natives such as bluebunch wheat-grass and Idaho fescue—to become established. The BLM was able to get ranchers to keep their cows away from the attractive, succulent growth. But they had no such luck with the horses. By the summer of 1981, sixty-five of them were camping on a section of new growth. Fecal analyses of wild horses in Idaho have shown that nothing gives as much gastronomic pleasure as bluebunch wheatgrass. According to Ben Garechana, the BLM's range conservationist in nearby Salmon, the horses denuded twelve hundred acres of new grasses.

Finally, in March 1983, after seven years of suits and appeals, a district court in the District of Columbia ruled in favor of the BLM. It decided that the bureau knew its business when it spoke of horse weights and eating habits, population growth rates and damage to the range. The court said that the BLM could reduce the horse population to any number between one hundred eighty-five and three hundred forty. That fall, the BLM rounded up three hundred eleven horses, nine of which were returned to the range because "they had good color in them." This left approximately two hundred-sixty wild horses in the Challis region, a figure theoretically low enough to necessitate no further roundups until 1986.

The roundup was conducted under the watchful eye of Susan West, one of Joan Blue's Washington-based "special assistants." Susan West had never seen a wild horse roundup before; her experience was limited to watching race track horses. Apparently, she was more-or-less satisfied with the BLM's treatment of the horses, with one major exception. She objected to the cowboys using prods to get the wild horses out of the corrals and into trucks. One wrangler at the corrals assured her that if he didn't make use of the charged red-and-white stick, they would be there for several days.

On October 4, 1983, the AHPA issued a press release on the round-up. It reported the court order, and then claimed that while the AHPA had had to compromise on their original design for the horses, the organization had nevertheless proven its point: that the area can support more wild horses than the BLM originally maintained. In 1976, the BLM had said it wanted to reduce the population to one hundred fifty, "plus an unspecified number of immature animals." The court order had set the minimum horse population at one hundred eighty-five.

Joan Blue was not the first person associated with the AHPA with whom I had a revealing face-to-face encounter. Nine months before going to Washington, I had paid a visit to Mark McGuire, Chief Investigator for the Nevada Humane Society and the state's principal representative for the AHPA in wild horse matters. In the latter capacity, his responsibility is to keep Joan Blue up to date on proposed roundups, on the condition of mustangs at Palomino Valley, and on policy changes under consideration by the Nevada BLM. McGuire confesses, however, that his humane society chores keep him so busy that he's able to spend only a "very small portion" of his time monitoring the BLM's wild horse program.

One of McGuire's major complaints is that the Nevada BLM's mustang population figures are too high. I asked him what kind of evidence he had to back up his assertion. He cited two instances, both short trips in which he had driven around in his four-wheel drive looking for horses. On one occasion, he went into the Clan Alpine Mountains, another time into Antelope Valley. "I saw few horses on these trips, nothing like what the BLM claims."

"Did you get away from the roads on foot or horseback?"

"No."

I noted that I too had had the same problem, and discovered that seeing them depends a lot on the time of day and a willingness to get away from the road and look for them. I pointed out that the BLM only records numbers of horses actually seen, and that careful studies done by the University of Minnesota had shown that BLM figures are understated by as much as fifty percent in heavily forested areas like the Clan Alpines. "Furthermore," I said, "the Nevada BLM doesn't take into account annual increases in the horse populations. Many of the population figures they use are four and five years old. No one that I know would claim that the horses aren't reproducing at a healthy rate or that the herds aren't growing rapidly. With a handful of exceptions, all of the one hundred twenty-five or so wild horse areas in the state are presently undercounted."

"Maybe that's right, what you're saying," he said. "I admit I don't follow that closely what's happening. I don't have the time." Then he added, "Joan Blue thinks that the number out there is overstated. She doesn't think the numbers are viable. We need more wild horses in Nevada."

"More than thirty-five or forty thousand?"

"That's how she feels."

I asked McGuire how his views on the wild horse issue meshed with hers.

"My main disagreement with Joan Blue is that she wants the total preservation of life. My top priority is the suffering. I'm interested in seeing animals put down humanely rather than going through life miserable. There's too much suffering among the wild horses."

"How would you solve the problem when there are too many horses in an area?"

"I'm not going to get into that."

I pressed on. "Can you be more specific about what you have in mind?"

"I'm for some kind of program of population control, that's the only alternative I can think of. I think some type of chemical birth control would be the answer. Money being used for capture and adoptions could be used for finding the right birth control method." There was a pause, then he added, "I'm not so much a preservationist as I am a humanitarian. Suffering is definitely the primary issue and it all comes down to eliminating that. What we all want comes down to how to get rid of that suffering. We want the horses to live in peace and harmony in their environment. That's my feeling."

"Is the AHPA humanitarian in your terms?"

"I see myself as a humanitarian," he repeated. "I can't speak for Joan Blue and the AHPA."

"Or Hope Ryden?"

"Or Hope Ryden."

Hope Ryden has written that she prefers to see horse populations regulated naturally rather than by human intervention. Hope Ryden has said, "Nature, if permitted to do so, would crop horses for their own good, removing the unfit, the stupid, the careless." She doesn't feel that one need worry much about damage to plant species from too many wild horses, for "other limiting factors such as stress and weather cycles cause the decline of a species long before its habitat is destroyed." As for those who might feel a bit squeamish when hearing what horses go through when they are victimized by the elements, she has said, "No doubt the public would have to be educated to accept the effects of winterkill, but in this the wild horse could serve an educational purpose."

Hope Ryden has been critical of the BLM because of its desire to maintain good-quality range by reducing mustang numbers. She has suggested that such protection is "artificial maintenance." Artificial it may be, but if wild horses were allowed to overgraze until some sort of "equilibrium" was reached, there would be even more severe desertification than currently exists. After grasses disappeared, horses would turn to forbes and brush, favored foods of deer and antelope. Numerous plant and animal species might be eliminated or isolated in small, barely viable pockets. Furthermore, as one Sierra Club member has asked in response to Ryden's position, if wild horses can overgraze until equilibrium with natural forces is achieved, should we not allow sheep and cattle to do so as well and forget our concern for wildlife, plants, and soil?

There is certainly a case to be made for allowing many animal populations to control themselves. But it is a difficult case to make for wild horses. They have no natural predators, they reproduce at a formidable rate, and they do considerable damage to rangelands. Ironically, one of the better examples to refute Ryden's opining on the matter comes from the very area she knows best—the Pryor Mountains. During the winter of 1969–1970, an independent sort by the name of Ron Hall observed horses in this topographically dissected area of sparse vegetation in southern Montana. Hall found that the horse herd had increased to about two hundred fifty animals that year, about twice the number the area could reasonably support. By late winter, a winter that was not particularly severe, the horses were pawing at the roots of plants that had already been grazed off. More than a dozen died, and many others were weak. Eventually, the government had to reduce the Pryor Mountain population by half.

In 1982, a National Research Council Committee on Wild and Free-Roaming Horses and Burros examined Ryden's thesis, and those of scientists, and it looked into cases like those in the Buffalo Hills and the Pryor Mountains. The committee concluded that wild horses do not self-limit their numbers before "increasing to the point of significant vegetation impact." Another report prepared by a number of distinguished biologists came to a similar conclusion, adding that "human intervention is required in management of animal numbers."

In 1979, the State of Nevada, on behalf of the Nevada Department of Wildlife, sued the BLM for mismanagement of its wild horse program. The state contended that the wild horses were consuming grasses and drinking water that belonged to wildlife, destroying cover for small game around water holes, and trampling riparian and flood-plain habitats, thereby impairing the ability of fish to spawn. To rectify matters, the Department of Wildlife wanted the gathering of wild horses put in the hands of the state and the horse population reduced to 1971 levels. The suit asked the court to declare that Nevada has the constitutional power and right to protect its wildlife, claiming that without more enlightened management of the horses, "a desirable ecological community could not be achieved and maintained."

George Tsukamoto, chief of Nevada's Division of Game, made it clear to me that he was not optimistic about the outcome of the lawsuit against the BLM. He said that it was unfortunate that no one in the Department of Wildlife had pushed for state control over the mustangs when proposals for a wild horse act began to gain momentum in the late 1960s. "If the state had done that, we wouldn't have had problems like we've had. The issues would have been localized and there would have been no national outcry over the horses. The problem was that all of the state's wildlife agencies were against the Wild Horse Act when it was passed in 1971. None of them wanted to take control because we're only responsible for resident wildlife species. Not migratory species, or feral animals."

Tsukamoto went on to say that no one thinks of wild horses as predators, "and yet they're very effective as predators on other animals. Horses aren't wildlife anymore than feral cats are wildlife, and cats are predators too. They're very effective at killing quail. Coyotes, skunks, and bobcats also kill quail, but they're natural predators. They're wildlife species. Horses aren't predators quite like feral cats, but horses rob wildlife of their resources. Antelope and deer need good winter range, and when there are too many horses they don't have it."

I asked Tsukamoto what kinds of studies the Department of Wildlife had done to substantiate its case against mustangs.

"We don't have to go out there to prove anything," he said. "We're not concerned in showing specifically the damage being done. We know that when you've got a large animal out there, in those great numbers, taking up that much living space, there are going to be problems. Something has to give for what all those horses are taking."

Then, sensitive to my request for documentation, he asked, "Do you know what happened to horses and wildlife in the Buffalo Hills?"

I said I did.

"You're probably also aware of what happened in the Owyhee Desert?"

I nodded.

Then he drew my attention to Jo Meeker's study in the Sheldon Antelope Range in northwest Nevada in the late 1970s. For her master's degree, she spent two summers studying wild horse and antelope diets. By examining the feces of both animals, she discovered that only thirteen percent of their diets were composed of the same plants. Only phlox, a forb, accounted for more than five percent of either animal's diet, and this was the second most common forb in the Sheldon Range. She recorded twenty-two distinct plant species in horse feces. When food is in short supply, however, the picture changes dramatically. Then there is considerable dietary overlap among all herbivores—elk, deer, antelope, horses, and domestic stock. At such times, all animals on the range are forced to eat the less nutritious parts of plants. Digestion is more difficult, nutrient content decreases, and animals—particularly those least able to compete—lose weight.

Not only do wild horses eat a lot, five to six times as much as a deer (even more compared to antelope), but the mere size of horses—six hundred to eight hundred pounds, a thousand pounds here and there in Nevada, more in Wyoming—means that they trample delicate vegetation. When their numbers are high the damage to plant and small animal life is considerable; in Nevada, horses have been particularly hard on young cottonwood trees. Grasses and forbes sustain the heaviest trampling damage. Ducks and sage grouse depend on sagebrush for nesting, they need perennial forbes, and they like to put their nests near water holes. But the commanding presence of horses around water, their destruction of plant life, and the muddying of water which prevents the growth of aquatic life make these sites unattractive to nest and shore birds alike. Whereas there may be two

182

or three pairs of waterfowl per acre of unmolested marsh, none can be seen when horses are abundant.

Several years ago, Bob Autenrieth, a wildlife research biologist with the Idaho Department of Fish and Game, conducted a study of antelope along the east fork of the Salmon River. He found that an overabundance of wild horses destroys protective cover and thereby makes fawns especially vulnerable to predators such as coyotes, bobcats, and golden eagles. Other antelope young with inadequate cover die from hypothermia, or they're abandoned by cautious mothers. Autenrieth also discovered that horses and antelope are neighborly with one another so long as there is plenty of water and vegetation for all. But when scarcity sets in, usually in late summer, antelope flee at the sight of horses.

Bighorn sheep are also victimized when there are too many horses around. Because of Stonewall Mountain's ideal topography—rough terrain and steep talus—the Nevada Department of Wildlife put a couple of dozen desert bighorn sheep there between 1975 and 1977. One of the aims was to reestablish the state's bighorn sheep population as it existed prior to 1850.

By 1984, the Stonewall Mountain bighorn sheep population had grown to one hundred. But according to Robert McQuivey of the Nevada Department of Wildlife and quite possibly the West's reigning expert on desert bighorn sheep, if something is not done about the wide-ranging and more adaptable wild horses that also utilize the mountain's vegetation, the sheep population may be lucky to reach one hundred fifty. At that point, McQuivey reckons, the sheep numbers may start to decrease; the population might even disappear. It will certainly not have one chance in one hundred of reaching five hundred, a good possibility were Stonewall Mountain not excessively utilized by mustangs. McQuivey maintains that it doesn't take a biologist or a well-funded government study to understand that with more than five hundred horses and substantial numbers of mule deer and chukar on Stonewall Mountain, there is certain to be overutilization and hedging of plants. It is equally certain that the horses will be the last to suffer. Unlike bighorn sheep, when horses' food supply declines they can move off the mountain's slopes onto the flatland without constant fear of becoming someone's trophy.

"I'm probably a typical Sierra Club person," Rose Strickland says. "I didn't know anything about horses before I came to this state. I knew more about burros; I ran into them down at the Grand Canyon. When I got to Nevada and found out how horses are involved in this whole public lands thing, my attitude radically changed."

Young, vibrant, forthright, opinionated, Rose Strickland is among the Sierra Club's most informed students on the West's public lands. Rose takes her activist vocation seriously. She hones her skills by taking classes in botany and zoology. From experience gained while working for HUD, she knows that facts alone are not enough to get her the audience the public lands issue demands. That's where Nevada's wild horses enter the picture. "As soon as I go before a Sierra Club group and talk about conserving public lands, my audience falls asleep. It's even worse if I try to talk to the public about overgrazing on our land. But boy, let me mention horses and I've got headlines in the *Reno Gazette.* Everybody has an opinion about wild horses. They hate them or love them, they want them for their own, they. . . . So horses are the key, my way of focusing attention on public lands issues." Which, to Rose, are dark and depressing matters. "I see the public lands issue in terms of animals who are suffering, agencies that do not do their jobs, livestock ranchers who are *totally* into making a living, and the devil take the hindmost."

Rose agrees with BLM policy in supporting a multiple-use concept for public lands: wildlife, wild horses, and cows should be able to coexist in harmony. Where she disagrees is on policy implementation, specifically on the fact that, to her way of thinking, there are far too many livestock on western ranges, and the BLM and its predecessor, the Grazing Service, are to blame. These agencies, she believes, have always been little more than fronts for ranchers and, as a result, "the public lands are beat to hell." She would like the BLM to make good on one of the avowed purposes of the 1934 Taylor Grazing Act, which was to "stop injury to the public grazing lands by preventing overgrazing." Rose won't say by how much she'd like cow herds reduced, but a figure of fifty or sixty percent would certainly bring forth a broad smile. On the other hand, she won't listen to an argument that all livestock should be removed. "Because then the empty spaces would be taken by MX missiles or by the navy. And that I *certainly* don't want." She has a similar attitude toward the usefulness of wild horses. "At first I was really resentful about the wild horse thing and the attention it was getting, but then I began wondering what the wild horse could do for the public lands. Of course, it can do a great deal—by keeping them open."

Rose Strickland is a formidable critic, and should anyone doubt her claims about the deplorable condition of the west's public lands, she'll cite chapter and paragraph and page number from any one of several environmental impact statements that clutter her home. She can, for example, point out that the Nevada Range Report of 1974 shows that the state's range is in poor to fair condition, and in god-awful condition around most riparian areas. If you don't believe her summaries of the bound documents, she'll command you to go to the library and check out one of many scientific reports that document increasing desertification of western rangelands.

One, prepared by a United Nations conference of scientists from one hundred nations convened in 1977, found that the most important cause of the modern expansion of deserts around the world is the grazing of domestic livestock. A 1981 report of the Council of Environmental Quality, entitled *Desertification of the United States,* agrees: " 'Improvident pasturage,' or 'overgrazing' as it has come to be known, has been the most potent desertification force, in terms of total acreage affected, within the United States." By the estimates of many scientists, more than two hundred million acres, most in the West, are being subjected to "severe desertification." Almost the entire state of Nevada is in the "moderate" category. Desertification, which saps the land's ability to support life, has numerous results: the reduction of surface waters, declining groundwater tables, unnaturally high soil erosion, salinization of topsoil and water, and the desolation of native vegetation.

Commenting on desertification, biologists Denzel and Nancy Ferguson write: "Today, desertification is gobbling up public rangelands because these lands continue to be grazed beyond their carrying capacity. The damage already done is appalling—many desert areas being only one-tenth as productive for livestock as they were when white men first came on the scene. Although modern stockmen attempt to blame all range deterioration on the era of the great stampede, indisputable evidence demonstrates that the damage is still going on. Perhaps the most criminal aspect of current desertification is that the federal government, by giving enormous subsidies to western stockmen, encourages them to continue their attempts to extract something of value from sick and degraded rangelands."

If none of this is convincing, or if you look like someone who requires humor and a gilded pill, then Rose Strickland will cheerfully tell you about her "cow pie test." She begins by revealing how much she likes to backpack into Nevada's out-of-the-way places. Then, smiling wryly, she asks, "Have you ever tried to set your tent up in the desert? If you did you probably had to kick several cow pies out of the way. After I had done this enough times I came up with my cow pie test. I decided that if there is more than one cow pie per square yard the area is overused."

Rose will admit that wild horses overgraze every bit as much as cows. The admission, however, is qualified and worked to advantage: "The reason the horses overgraze is that that land out there is the only place they have left. There are too many cows, and my goal is to get the agencies to take some of them off. My position is that horses have as much legitimate right to the public land as livestock or recreation or anything else. The official BLM position is that we don't have enough data to reduce the numbers of cows, so they refuse to do anything. Then the BLM says that if there's overgrazing five to ten years down the line, that will be the time to bring down the number of cows. But with that logic they're being caught on their own petard, for if that's true for cattle, why isn't it true for horses? Do they have any special data for horses that they don't have for cows? No, they sure don't. That the BLM has the data to reduce horses but not to reduce cows— as they say—seems fairly inconsistent to me. They won't go for my paying for the horses like ranchers pay for their cows because the real problem is that the horses compete with cows. You have to get rid of the horses is how they see it." Rose feels that the government is doing her a considerable injustice by not accepting her $1.40 monthly checks—or whatever the yearly rate happens to be—to keep a wild horse on public land, the same amount the government accepts from ranchers for each grazing cow. She returns to the current monthly figure for grazing a cow with the regularity of a cuckoo slamming through its doors four times an hour.

Rose Strickland's logic is many-sided, and she can be relentless. She asks, "How do you know there are too many horses in an area when the BLM says there are too many?" Answering her own question, she says, "Because the BLM tells you the horses are overgrazing in that area. They never tell you about the cows. They have the information. There have been several environmental impact statements in the last five years and they cost millions of dollars, but the BLM doesn't like the results. The results show that there are too many cows and the land is overgrazed. They never tell you about all the areas that are overgrazed where there are no horses."

Nothing will convince Rose that a rancher will voluntarily reduce the number of cows he is grazing, even when he sees the range deteriorating. To her, any long-run self-interest theory in this regard is a lot of "Bullshit," which she spells with a capital B. In fact, as she sees it, ranchers will always have as many cows as possible. She reasons that many ranchers have bank loans to pay off and the only way they can get the necessary income is to run as many cows as they can get away with. Or, if ranchers want additional loans, they need to convince the bank that they can repay them and the best way to do this is to have large herds. Pounds on the hoof are the measure of ability to repay debt.

Rose's uncompromising case is supported by many people, including some within the BLM. George High, who works law enforcement in Nevada for the BLM and is constantly in touch with ranchers, says that some cattlemen will run half again as many cows as they have a right to by law. In his words, this philosophy reduces to: "Get everything you can." High says he has known old-timers who claim that when they were kids they frequently saw grass on Nevada's ranges as high as a horse's belly. Now in those same places it's no higher than a coffee cup.

There is some evidence to support this. In *A Range History of Nevada*, Ben Hazeltine, Charles Saulisberry, and Harry Taylor state that, "A significant feature of the virgin sagebrush-grass range [the most widespread vegetative type in Nevada] was the abundance of palatable grasses and weeds that grew under and between the shrubs. Principal grasses were wheat grasses, Idaho fescue, needlegrasses, wild rye, and Indian ricegrass. Within the sagebrush-grass range there occurred large areas of saline-alkali bottomlands on which wild rye is said to have formed almost pure stands with only very scattered sagebrush plants in evidence." These were ideal winter pastures. In one valley in northwestern Nevada, one nineteenth-century rancher is alleged to have wintered between ten thousand and twelve thousand head of cattle on rye grass.

While there seems to be general agreement that Nevada's range is not nearly what it once was, and that one now sees a good deal more sagebrush where once palatable grasses predominated, the journal entries of early travelers suggest considerable variation within the state. Horace Greeley, in his overland journey from New York to San Francisco in the summer of 1859, wrote (as he followed the Humboldt River): "I thought I had seen barrenness before—on the upper course of the Republican, on the North Platte, Green River, etc.—but it was green, if the regions washed by those streams were not. On the above-named rivers, I regarded cottonwood with contempt; here, a belt, even the narrowest fringe of cottonwood would make a comparative Eden. The sagebrush and greasewood, which cover the high, parched plain on either side of the river's bottom, seem thinly set, with broad spaces of naked, shining, glaring, clay between them." In central Nevada, government expeditions found grass on high-elevation slopes, in mountain canyons, and in moist valley bottoms. But elsewhere, bunchgrass was scattered. In 1859, J. Schiel reported: "The entire country between the Humboldt Mountains and the Sierra Nevada of California is a rugged mountain area with only the most miserable vegetation. Only here and there on the slope of a mountain or in the narrow valleys and ravines can one find grass for the animals." In the 1870s, a government geologist spoke of the "everlasting sagebrush! This is by far the most prevalent of all species, covering valleys and foothills in broad stretches farther than the eye can reach, the growth never so dense as to seriously obstruct the way, but very uniform over large surfaces, very rarely reaching the saddle height of a mule."

But don't ranchers want to improve the range, if only to be able to run more cows and make more money? Rose's answer is succinct: "They want to improve it all right, with federal money. In Nevada, improving the range means planting crested wheatgrass in every valley in the state so they can run still more cows. All the BLM thinks about is its benefit–cost ratios, the market value of its projects. Forage production for cattlemen is the only thing on their minds. They don't even think about problems like all the grasshoppers you get with crested wheatgrass, or the loss of soil nutrients that keep out our native species."

Are there other problems with crested wheatgrass? Several, it seems. One of the most notable is that it has been planted by the BLM at public expense to the total exclusion of other species, and this has meant a loss in the diversity of life found within plant communities. It is a generally well accepted ecological principle that ecosystems with many different species tend to be stable. For example, because diseases are usually selective, diverse communities can withstand a severe attack. Not so with a monocultural system—a field planted exclusively with an exotic like crested wheatgrass. Furthermore, even where there are numerous other species alongside crested wheatgrass, many may be lost over time. Most plants are stressed, or in trouble, if thirty to fifty percent of the above-ground portion is eliminated. But crested wheatgrass can withstand almost double this amount of cropping—up to eighty percent. As a general rule, the more intense the grazing in an area, the greater the loss in diversity.

A loss in the diversity of plant life may have negative consequences for wildlife. Snakes fare poorly in areas planted with crested wheatgrass, and lizards become scarce and may even disappear. The density of nesting and non-nesting birds and their species diversity are often low, much lower than for areas in which overgrazed sagebrush predominates. Because the presence of crested wheatgrass encourages greater grazing pressure by ranchers, there is greater likelihood of soil loss, no small matter. Nature requires hundreds of years to replace it. Little wonder that conservationists such as Rose Strickland have been opposed to seeding crested wheatgrass, and have described these exotic monocultures as "biological deserts."

Rose likes to note that the money ranchers pay to keep a cow on public land for one month, known as an AUM (animal unit per month), is not only miniscule—$1.86 in 1982, $1.40 in 1983, $1.37 in 1984—but that they get a large chunk of it back in government-financed range improvements. Fifty percent of grazing fees collected by the BLM goes for range improvements in the district where they are collected. Most of the rest is turned over to the U.S. Treasury.

How does Rose explain the fact that in many areas of Nevada ranchers have fewer cows on the range than permitted by the BLM? That's also simple, to her way of thinking: The land cannot support the number of cows they would like to run, the market for beef cattle in recent years has not been good, and you cannot keep cows on the land during much of the winter because you need hay, which means water, which means energy, and energy to pump water is the most expensive thing there is in northern and central Nevada. Rose does not accept arguments that ranchers have reduced their cow herds because grasses are being eaten by wild horses. She points out that ranchers periodically reduced their herds prior to 1971 and the beginning of the wild horse problem, and that they did so almost solely because of low prices for beef.

The picture is always more complicated than hard-line advocates like Rose Strickland are willing to concede. For example, in only one of a score of Nevada counties—Elko—are there as many cattle today as there were forty years ago. The average age of people involved in Nevada's livestock industry is above fifty, and increasingly many ranchers work only part-time. In the Carson City BLM district, there has been a dramatic reduction in the number of cattle, and not at the behest of the bureau, or because of market prices, or because of food- and sex-hungry mustangs. Of fifty-five ranchers with permits to use federal lands for cattle in the Carson City area, only eight can be considered full-time, viable outfits. The situation is even worse for sheep ranchers, who have a very difficult time recruiting herders. Ranching is no longer seen as romantic by young people. Many have decided that life in the city is less demanding, less lonely, more fun, and a lot less uncertain.

Rose has a solution of sorts for improving the range. She wants to see both horses and cows removed. To her it's a swap game: "You can take a wild horse off the public land, but I want to see one cow also removed for each horse that goes." Since both horses and cows are to blame for the poor condition of the range, this strikes her as the only reasonable approach to the problem.

It even strikes some in the BLM the same way—some of the time. Milt Frei, who calls most of the shots on managing and rounding up Nevada's wild horses, agrees with Rose. "What I'm really for is good resource management," he says. "What I do is play horses against cows. Cows are primarily detrimental to range management. Let's trade cows for horses." He's also said, "I fight for horses because they're underdogs and have been so badmouthed. I see the FAA and the BLM as similar in subsidies they give to their respective industries. The FAA likes airplanes, we like livestock. BLM employees like the western cowboy image; they like ranchers. I like them too, but that doesn't mean you're necessarily going to give them everything there is out there. There are twenty-two thousand people in the livestock industry in the West and they control millions and millions of acres." On still another occasion, in answer to my question, "How many wild horses would you like to see in Nevada," Frei said, "As many as I can get! I'm an empire builder. I can do projects for the horses—put in water, etc. It would be up to managers to slow me down. I want as many horses and as few cows as I can get, commensurate with good range. We get paid to be program advocates."

Andy Anderson, a thoughtful range and wild horse specialist for the BLM in Carson City, believes that any talk about a cows-for-horses swap ignores a few basic facts. He points out that some areas have horses but no cows, some have cows but no horses, and others have both. "But in much of the West, the cattle population has nothing to do with horses and therefore in many areas of Nevada or elsewhere in the West, horse and cow swaps don't even make sense. Furthermore, none of these people who want to swap cows for horses talks about going out and removing wildlife. From my perspective, it's resources that matter. It's okay to have eight million horses in Montana or anywhere else if the carrying capacity is there." Anderson is concerned that the situation is getting worse by the day. He says, "If present trends continue, we'll be hip deep in horses, and livestock operators will soon be out of business." Yet his solution is familiar: reduce both the numbers of cows and horses, since where they do use the same ranges there are usually too many of both. Of course Anderson has his own prejudices. "I would rather see cows than horses on the range, because that's my background." Andy Anderson has a degree in animal science, the art of maximizing meat production for human consumption.

Though Rose Strickland and other Sierra Club members have been actively involved with wild horse issues for several years, there is still no national Sierra Club position on how they ought to be handled. The wild horse presents a sticky problem. It is considered an exotic rather than a native species and therefore does not fall under the umbrella of the club's National Wildlife Policy, often hailed as one of the most enlightened and biologically sound doctrines of wildlife protection. Thus far, the policy-making arm of the Sierra Club has not been persuaded by the argument of some wildlife professionals who contend that the wild horse is, in fact, a native.

By this line of reasoning, the wild horse, like the elk, arrived in North America approximately forty thousand years ago, coming from Asia via the Bering land bridge. Like other species which came at about the same time, the horse found a viable niche within North America's ecosystem and survived, until it mysteriously disappeared some eight thousand to ten thousand years ago. The cause cannot be attributed to competition with other animals; or a drastic change in the environment—horses died out everywhere in North America; or an unusually effective predator—such as sabertooth cats or Indians; but perhaps, as George Gaylord Simpson has suggested, in a fly-borne epidemic such as sleeping sickness. Whatever lies behind one of the more intriguing episodes in animal evolution, the niche which allowed the horse to survive for tens of thousands of years remained, according to the "native species" thesis. From the time the horse first made its way into North America until the present, there has been no appreciable change in climatic conditions or vegetation, at least not changes of the sort that eliminated a place for the horse within nature's North American scheme of things. Thus, when domesticated horses were reintroduced by Spanish conquistadores in the sixteenth century, and some escaped and became wild, they simply made use of an ecological niche that their Asian ancestors had exploited to good advantage. The wild horse, as it were, was merely returning to an ancient home.

There are dissenters to this reconstruction of a sketchy past. The National Academy of Sciences, in a 1983 report on wild horses, concluded that while much of equid evolution occurred in North America and species of *Equus* were found in America as recently as the Pleistocene, some eight to twelve thousand years ago, today's wild horses are not, in fact, occupying an otherwise vacant ecological niche. Apart from climatic changes, North American biota have changed profoundly since the horse disappeared. Intermountain vegetation is different from that which prevailed during the Pleistocene, different even from that encountered by European settlers, and many of the large herbivores that preyed upon horses or competed with them are no longer around. Thus, it cannot be argued that ecological niches that date back ten thousand years or more still exist.

Sierra Club national policy makers applaud this conclusion, though their reason seems to have as much to do with political considerations as with the reasoned facts of science. When the issue of how wild horses ought to be viewed was raised at a Sierra Club forum in 1974, one of its national directors, Robert Hughes, insisted that the club not develop a national position on the animal. He was, he said, concerned that by so doing the organization would be going against the meaning and intent of its National Wildlife Policy.

I asked one western Sierra Club member familiar with wild horse issues if there were other reasons why the national organization has not taken a stand. He said that there were deep divisions within the organization. As many as half of the club's eastern membership belongs to staunch protectionist groups, such as the American Horse Protection Association. Others simply feel that the quality of the range and the welfare of native wildlife take precedence over wild horses. Many western Sierra Club members are different in at least one major respect from their eastern counterparts. Familiar with ranchers, some feel that, up to a point, ranchers have a good case against the actively reproducing horses. Out of such tenuous generalizations as these, the Sierra Club's national directors seem to have come to the conclusion that the prudent thing to do was to look the other way. By so doing, the club was not likely to lose support on other issues deemed more critical. As one member noted, "The Sierra Club doesn't feel that the horse issue is a major one. There are at least ten that are more important. Take New York, for example, which has the second largest Sierra Club membership in the country. Air pollution, water pollution, energy . . . there's all kinds of issues making claims on their urban minds. My feeling is the Sierra Club will never take a position on the horses, not nationally it won't. Why develop one that's going to have problems?"

Irreconcilable differences there may be within the Sierra Club, but one chapter has not shied away from the issue. This is the Toiyabe Chapter of Nevada and eastern California. Within its geographical boundaries can be found the greatest number of wild horses anywhere in the country. One Toiyabe member who several years ago took the lead on the wild horse issue was Tina Nappe.

A mother and full-time employee of Northern Nevada Job Training Program for Washoe County in Reno, Tina over the years has found time to serve as a member of the Nevada State Wildlife Commission and on several BLM planning groups in which removal of wild horses was a central issue. For one three-year stretch, she put out ten issues a year of *Toiyabe Trails*, the chapter newspaper, and she personally wrote several articles on the wild horse: on its role in the settlement of Nevada; on the intricacies of the federal law; and on what the rancher's perspective has to offer conservationists.

Tina points out that while most of Nevada's wildlife is on public lands, most of the state's water is in private hands. Conservationists, in their anti-rancher rancor, have forgotten this simple fact about water, and its necessary implication: water is critical to the welfare of wildlife. By the very nature of their demands for lots of acreage on which to graze their cows, ranchers are actually protecting public lands. Get rid of the western rancher and it won't be long before private open spaces are sold to developers. And then, without water and riparian resources, many wildlife species will be greatly reduced in number. Or eliminated.

In 1974, when the secretaries of Interior and Agriculture proposed an amendment to the wild horse act that would allow the BLM to use helicopters on roundups, Tina was at the forefront in generating debate on the issue. Along with the club's conservation chairman, Tina felt that the amendment was necessary to prevent further range deterioration. By that time, there was already evidence that horse numbers were multiplying beyond anyone's expectations, and that the BLM was underfunded for the task of managing them. Tina was concerned for species diversity, "little-known plant and animal species," and she pushed her belief that "horses are not 'natural' to the Great Basin." Tina was—and is—concerned that too many horses or too much livestock will simplify nature's scheme of things, reducing both the number and variety of wildlife species. At the time of the proposed roundup legislation, she found herself at odds with Wild Horse Annie, who feared that helicopter roundups would be inhumane and that with more efficient gatherings, mustangs would be going to slaughterhouses. The adoption program was still in its nascency. The amendment that allowed helicopters to be used in roundups became law in 1976, and Wild Horse Annie's concerns proved unfounded.

Tina has worked hard to correct misconceptions actively entertained by wild horse activists. She has, for example, taken issue in print with Hope Ryden and Paulette Nenner, the editor of *Wildlife Involvement News*, a Sierra Club publication that provides excellent coverage on what is happening to wildlife around the world. Ryden has contended that, "A great number of wild horse herds exist in . . . harsh and inhospitable places where the concept of multiple-use has no practical meaning," while Nenner has asked: "Why have the wild horses been so heavily controlled when much of their habitat exists in semi-barren rocky land that is inhospitable for cattle as well as wildlife?" Tina argues that those familiar with wild horses in the Great Basin cannot recall instances in which their ranges did not overlap with livestock, and if horses use "inhospitable" areas, they do so only on a temporary basis. Tina, her mind's eye always fixed on the broadest possible concept of land—"Land is my love and my religion," she has said—has asked of Hope Ryden: "Is her definition confined to game species, livestock, and wild horses? Should it not also include a diversity of species including nongame wildlife and plants, unique ecosystems, and watershed protection?"

Tina, in her polite and engaging way, has pointed out that the fragility of the Great Basin desert is not conducive to intensive use or large numbers of grazing animals. The first explorers to come into the Great Basin saw few deer; antelope, though present in Nevada, did not exist in large numbers. Bighorn sheep, which use valleys and lowlands primarily in winter, numbered no more than about ten thousand when Caucasians first appeared. Today, there are five thousand bighorn sheep in Nevada.

Tina has been instrumental in developing a Toiyabe Chapter position that at once addresses major wild horse issues and yet does not compromise national Sierra Club policy. In her own words, "The Toiyabe Chapter's position on wild horses is to actively support continuance of wild horses on the range according to the land's carrying capacity and in consideration of other rangeland users such as wildlife, livestock, endemic plants, and unique ecosystems." She has urged the Sierra Club to aggressively support wild horses on public lands because they have aesthetic and symbolic value, and because a philosophy that emphasizes diversity means that there is a place in nature for all forms of life.

Under Tina's leadership, the Toiyabe Chapter supports BLM land-use planning and the involvement of citizen committees in deciding whether horses should be reduced in certain areas. Along with cattlemen and BLMers, Tina has sat on several of these committees, and though results are often slow in coming, she is a firm believer in the art of persuasion and reasoned compromise. She believes that BLM advisory boards and committees set up to deal with range management—the allocation of feed among cows, horses, and wildlife—have created an atmosphere of friendship and goodwill among interest groups. The committees have also given the BLM a chance to break with old ways. Tina says, "When the wild horse legislation came along, the BLM had mixed feelings about it. Half the people in the BLM at that time were range scientists and they looked at wild horses the way ranchers looked at them: they certainly are nice, but if they're not of economic value, then what are they really good for? The BLM has changed its attitude."

The committee approach to range management—Coordinated Resource Management Planning, it is called—has produced some ironies and turnabouts. In the beginning, cattlemen favored the broad-based committees, believing they would get very significant removals of wild horses in certain areas. They did, in a few cases. By the early 1980s, conservationists and wild horse activists had approved the roundup of more than a thousand horses in the state, and they also made some areas with obvious resource damage a high priority for removal of mustangs. But the gains have not been enough for the Nevada Cattleman's Association, and now the organization seems to want little to do with the committee process. On the other hand, conservationists like Tina, who were initially skeptical, believe that slow and cumbersome though committee work is, it is the only way to reduce the tension between ranchers, the BLM, and activist groups.

The worth of these kinds of committees granted, Tina would add that one committee she is quite familiar with has irked and baffled her. This is a state committee charged with finding an appropriate use for the Leo Heil Fund. Leo Heil was a wealthy Californian who may never have seen more than a handful of wild horses in his life; his knowledge of them probably came from newsprint. Nonetheless, when he died in 1972 he bequeathed $500,000 to the state of Nevada "for the preservation of wild horses in Nevada." A committee, headed by Senator Floyd Lamb from Las Vegas, was appointed by then-governor Mike O'Callaghan to decide how best to fulfill Heil's intentions. For more than a decade, the best the committee seemed able or willing to do was spend $16,000 to decide what to do with the money. Among proposals considered were: the establishment of a "Heil Visitor's Center"; a national wild horse museum at Palomino Valley and a nearby free-roaming herd of horses for the benefit of visitors; a reserve for mustangs with primitive markings in northern Nevada's Little Owyhee Desert; a ranch for unadoptable horses; the development of wells, pipelines, and reservoirs for the exclusive use of horses; a fund for monitoring BLM roundups to see that they are humane; the purchase of grazing rights for horses; support of legislation in Congress which would give the BLM authority to sell excess mustangs;

and an endowed chair at the University of Nevada for research on wild horses. This last idea was apparently one dear to Wild Horse Annie.

While it is difficult to determine the precise reasons for inaction on the Leo Heil Fund, the number of cattlemen on the committee and the fact that Floyd Lamb, whose sympathies and voting record lay with livestock interests, was the committee's chairman, offer possible explanations. While Lamb has said that he likes horses, he has also said that he just doesn't like too many of them. Over the years, reasons given for keeping the money locked up included the "ambiguity" of Heil's intentions, Lamb's opinion that there was no need for more research on wild horses, and his belief that there were so many horses that it was difficult to know what to do with the money.

In 1983 Nevada's governor, Richard Bryan, removed Floyd Lamb as chairman of the committee and expanded its membership from five to nine. New members included several representatives of groups long active on behalf of wild horses. In its first meetings it was clear that just about everyone had an idea how the money ought to be used. It also seemed clear that if cattlemen had their way, the money—approaching a million dollars in a Nevada Department of Natural Resources and Conservation bank account—would either be used to reduce horse numbers, or wouldn't be spent. Finally, in October 1984, after holding hearings around the state, the committee came up with its final recommendations. These included appointment of an impartial individual to carry out Leo Heil's wishes; creation of a trust fund for receiving tax deductible contributions for the state's wild horse program; the amendment of state legislation to include wild horses under present anti-cruelty laws; the collection of information on the history of the wild horse in Nevada and on the adoption program and methods for caring for mustangs; the coordination of research in Nevada and elsewhere on horses; and the establishment of two reserves in Nevada, one in the northern part of the state, the other in the south. The purpose of these reserves would be to attract those who want to see mustangs as well as those who wish to study them.

For many years, Paul Bottari, executive director of the Nevada Cattleman's Association, has been working to get the state's wild horse population reduced to what it was in 1971, "except in those areas where they were already excessive." By his calculations, the ideal figure would not exceed eight thousand. Like most of the cattlemen that he represents, Bottari would like to see horses that cannot be adopted out by the BLM sold on the open market. Bottari likes to point out that the National Wildlife Federation, the American Forestry Association, the Audubon Society, Nevada activists in the Sierra Club, Nevada's Committee on Natural Resources, Environment and Agriculture, and the National Wild Horse Organization support this position. As for those who strongly disagree with him, those he refers to as "extremist wild horse protection groups," he offers a challenge: "Come out and see the actual damage that the wild horses are causing. We would be happy to host such a tour."

From his Elko headquarters, Bottari frequently writes editorials and articles for newspapers, trying to present the cattleman's point of view. Nothing makes him more angry than journalists who have two hours to get the whole story on the West's wild horse problem, and then only hear it from one point of view. Bottari's battle is not easy, for even in much of the West, cattlemen are seen as plunderers of the public domain who have little sympathy for wild horses. Bottari and his constituency flatly deny the validity of this characterization. In their defense, they emphasize four points: ranchers put the first horses on the range and therefore the animals are private property; they have always found pleasure in seeing mustangs running free and, except for a small handful of ranchers, wouldn't want to see all of them eliminated; if they had wanted to get rid of all the horses they could have done so prior to passage of the 1971 wild horse act; and now, because wild horse numbers are large and growing, their very livelihoods are threatened. It is, they say, tough enough trying to make a marginal, middle-class income as a Nevada rancher, and if the horses continue to multiply as they have in recent years it will be impossible. Unfavorable press swells their anxieties. A few ranchers are nearly paranoid about what they see as the real meaning of the wild horse issue: it is an abominable scheme by conservationists or environmentalists—"whatever you call them"—to gain control of the public lands, and for no reason other than to get rid of the rancher.

Bottari is anxious to point out that wild horse numbers in the state have been increasing while the cattle population has been decreasing. He notes, for example, that there were 366,000 head of cattle and domestic horses using BLM lands in 1971 in Nevada, but by 1979 the comparable figure had dropped to 338,000. Meanwhile the wild horse population had increased more than threefold. "The livestock industry seems to have been the only one giving. If the increasing number of wild horses aren't the major cause of range deterioration in the state of Nevada, I'll eat my hat."

"But why have cattlemen reduced their herd numbers?" I asked him.

"Livestock operators have taken the voluntary reductions in order to save the range from excessive wild horse damage."

Yet when I asked Bottari what role fluctuating demand and changing prices for beef have had, he admitted that it is difficult to ascertain what proportion of herd reductions reflects lower market prices for beef and what proportion represents genuine concern for the quality of the range.

In one flier distributed by the Nevada Cattleman's Association, Bottari claims that in the first ten years of the wild horse program, the bill to the American taxpayer exceeded $200 million. In fact, the amount appropriated by Congress during this period was approximately $35 million. Of course, if all kinds of elusive costs were to be included in Bottari's figures, especially those for damage done to the range, his figure may be conservative indeed.

Whatever the real cost of the horse program, Bottari is eager to have it pay for itself—another rationale for selling excess wild horses to the highest bidder. A problem with this argument is that it highlights a double standard. In 1981, when there was a severe drought, the Department of Agriculture's Emergency Feed Program gave two hundred thirty-eight ranchers in Nevada and in one adjoining California county four million dollars for livestock feed. Leading the way in assistance received was Elko County: ninety-five ranchers got $1.85 million. Four other Nevada counties, all with sizeable wild horse populations, together received $1.82 million.

The lean, intense Bottari sees his battles with the BLM and conservationists and wild horse activists as an endless string of Sisyphean frustrations, made all the more difficult because he receives very little support from the National Cattleman's Association. The mother organization feels that it has more important issues than the disquiet of Nevada ranchers: cows fattened on the West's public lands represent a miniscule fraction of all beef consumed by Americans.

Aware that his effort alone was not enough, in 1981 Bottari hired a Carson City consulting firm called Resource Concepts. Resource Concepts advertises itself as a multi-disciplinary team of range managers, civil engineers, hydrologists, and agricultural economists who specialize in resource planning, environmental analysis, and water rights. Since the company opened its doors in 1978 in an attractive two-story clapboard house across the street from Senator Paul Laxault's law firm, it has prepared numerous environmental impact statements on behalf of ranchers. By the early 1980s, Resource Concepts was representing four of the state's six grazing boards. Partners in the firm will quickly admit that they are perceived to represent the livestock industry. But John McClain, one of its founders and a former range and soil specialist for the Soil Conservation Service in Idaho, will just as quickly add, "We look at things more objectively than others do."

One of McClain's proudest achievements since going into the consulting business was evaluation of an environmental impact statement prepared by the BLM. In 1980, the Nevada Department of Agriculture, at the request of the Nevada Cattleman's Association and the governor, asked Resource Concepts to reexamine new, and lower, cattle-stocking rates established by the BLM for some three and a half million acres in southeastern Nevada—the Caliente Resource Area. It was claimed that if the BLM's new figures were used, a "minimum of five medium-sized cattle operators and twenty-two small cattle operators would be forced out of business," and, in addition, "implementation would cause substantial social adjustments among families within the area." The task defined, McClain and a team of field workers did ocular reconnaissance and transects of vegetation types. They carefully identified and weighed all plants on the transects. When the field study was finished and they had checked the BLM report, they discovered that the BLM had made several measurement and computational errors. McClain was able to convince the Department of Agriculture and the governor that "the BLM's Caliente survey was invalid and no changes in livestock stocking levels should be implemented." He also argued "that the practice of basing forage allocation on range surveys be discontinued in lieu of a program to monitor utilization and range conditions."

Milt Frei feels that McClain's Caliente study had devastating implications for the BLM and the wild horse program alike. "It forced the BLM to admit that it doesn't know how much forage is on the range. It set the bureau back a hundred years. Now we're not using range surveys and we're making none. The official bureau position is: we don't know what's out there for horses and cows." In fact, the BLM hasn't had a good feel for the carrying capacity of its rangelands for more than two decades, at least since the early 1960s when the last surveys were made.

Frei also says that the Caliente study greatly dampened the enthusiasm that many ranchers had for the Sagebrush Rebellion, a movement to put the West's federal lands in the hands of states. The rebellion began in 1977 when the Nevada legislature decided to challenge the federal government's control of forty-seven million acres, roughly eighty-six percent of the entire state. Although most Nevada legislators have been in favor of the Sagebrush Rebellion at one time or another, and President Reagan has been for "privatization," or sale of the public domain, it is highly unlikely that Nevada and other western states can overcome legal precedents that favor the federal government. Furthermore, some legislators, and some ranchers—even those with severe wild horse problems—appreciate that a successful rebellion would spell disaster. The BLM spends millions more on Nevada's public lands than it collects in grazing fees, especially for the rehabilitation of overgrazed rangeland. With a successful rebellion, there might well be quick solutions for ranchers with wild horse problems, but at a price that few would want to bear once they sobered up.

When McClain was approached by Bottari to present the cattleman's point of view on the wild horses, he took a personal interest in the job. He wasn't interested, he says, in doing a "study." Rather, what was needed, in his opinion, was "an educational and informational program on the horses." He felt that the state has far too many horses, that they're damaging precious range resources, and that something had to be done about "subsidizing wrong attitudes and prejudices." He decided that the best format would be a twenty-five-minute slide show with a full sound track. The final product would be based on "a limited reading of the wild horse literature" and interviews with two cattle ranchers and two BLMers. The Nevada Cattleman's Association didn't have enough money (and it wasn't thought necessary) to demonstrate wild horse damage.

Shortly after we met, McClain invited me to attend a Rotary–Lions Club breakfast meeting at Sharkey's Casino in Gardnerville to see his slide show, "The Trouble with Mustangs." By this time, he had already given the presentation in Washington, D.C., and to a number of local professional gatherings around the state. He was boyishly proud of his effort and, for all its limitations and simplifications of a complex issue, it is an engaging show.

"The Trouble with Mustangs" opens with a helicopter chasing wild horses to the tune of a strumming guitar. This dissolves to a shot of an underfed mustang and the claim that wild horse numbers are growing at twenty-percent a year, a figure too high by double. Then the confident, deep-voiced narrator summarizes the horse's history in the Americas, noting that those on Nevada's ranges can be traced to the nineteenth century when ranchers turned them out to graze beside their livestock on unfenced lands. "The horse was the internal combustion engine of the nineteenth century," the narrator booms. Then, he says, "The horse numbers increased in the dry years of the late twenties and the early thirties when people left their homesteads for work in the city." This was a time when "the only controlling factors were nature, mustangers, and the ranchers, who attempted to keep the herds in balance with the carrying capacity of the range." The rancher, one senses from the McClain portrayal, was a well-meaning conservationist.

"The Trouble with Mustangs," the narrator continues, began with passage of the 1971 Wild Horse Act. A band of proud mustangs appears on a parched hill, then this slide dissolves into a shot of a Carson City BLM wild horse expert. Poised and pleasant, he says: "The 1971 Act was literally shoved down the throats of Congress because of the tremendous letter writing of the school children, and the very, very emotional attitude of all the people. We got letters, and letters, and letters from a whole class as if the teacher had given the children exactly what they should say and the letters vary . . . say in twenty letters. They varied just one or two words in them, and basically 'please don't kill the horses' and that sort of thing, and so there was no way that any congressman or senator wouldn't vote for the act."

As the slide show comes to a close, the narrator says that three changes in the federal wild horse law are necessary: (1) the BLM must be given commercial sale authority to dispose of unadoptable feral horses; (2) revenues from the sale of these horses should be used to make the management of the herds self-supporting; and (3) the BLM must be given a clear mandate to manage the horses, something that cattlemen and those in the BLM feel was never provided for in the 1971 law.

The visual portion of the show over, McClain entertained questions.

"What is being done?" one short-sleeved, balding businessman asked.

"What can we do?" said another.

After McClain confidently answered several questions, Matt Benson, a longtime local ranch foreman, got up and waved a petition. He made his own pitch for new laws regulating the horses. He said that there were "too damn many horses" and "their numbers ought to be limited to the number the range can support. You have to manage them as we manage other species in the realm of food supply." He said that unadoptable horses should be sold and the revenue returned to the BLM to subsidize the program. Then he said, "Sign this petition if you believe in what you saw today and the need for something to be done. This will be sent to your state and national congressmen." The petition began circulating, but before the ink was dry on the first signature, one of the club's members went to the front of the room and got down to the serious business of cajoling those present into buying tickets to a chicken stampede.

The chicken stampede, I learned, came off with great fanfare and produced many a belly laugh. The petition, however, disappeared somewhere within the labyrinthine depths of government bureaucracy. The principal effect of this petition and others, I also learned, sharpened John McClain's softer edges. Two years after our meeting, I began hearing from a variety of sources that McClain was telling anyone who would listen that the only way to beat the BLM and the horse problem was to bring a lawsuit.

Roger McCormack is Nevada's associate state director for the BLM. More the executive in color and manner than other BLM bosses, he seemed to welcome my inquisitiveness about the wild horse program. Perhaps it was the opportunity to tell me, as he has others, that the issue takes a disproportionate amount of his time, and that no other single BLM concern is so tough to deal with.

McCormack feels that the best way to judge whether or not the BLM is doing the right thing by the horses is to see if everybody—cattlemen, preservationists, humanitarians, legislators, wildlife administrators, the public, you name them—is equally mad at the BLM. "Then we're probably pretty close to what we should be doing. I've been in this business quite a while and I've learned there are very few black and white decisions. Nowhere is this more true than when it comes to the horses."

McCormack seems to have adopted the philosophy of his previous boss, E. I. Rowland, state director of the Nevada BLM in the mid-1970s. In Rowland's words, "We can't make anybody happy. We can't lock up all the land for recreation. We can't give it all to the livestock people or the wild horse people. We can't give it all to the environmentalists or the off-road vehicle people. We have to try to accommodate everyone in a balanced way. And nobody is satisfied."

McCormack explained that the BLM is now more highly politicized than ever, and the blame belongs to former President Jimmy Carter, who reformed the Civil Service. Carter made it easy for BLM administrators to be fired if someone above them thinks they've gotten out of line or are making "improper" decisions. "Now, administrators are continually looking over their shoulders. Everybody's trying to make sure they don't stand out in decision-making. You gotta be a tightrope walker to survive."

And survival in this sense means being unusually sensitive to anyone who can make trouble. It means responding to the mere perception of a threat. If wild horse activists such as Dawn Lappin are dissatisfied with a proposed roundup or a decision that they believe adversely affects the horses, they can get quick access to Nevada's state BLM director. He'll listen patiently, and he'll follow the logic, and if he can't come up with better logic and satisfying answers he'll get on the phone to his district manager.

I asked McCormack if he could be specific about why he and other BLM administrators wish that the mustangs would become someone else's headache.

Point 1: Shrill sensationalism, McCormack maintains, beclouds everything the BLM tries to do with the horses. Grammar school facts and inelastic emotions ride roughshod in the press and in the public's mind, and, government bad guy that the BLM is, the best that one can hope for is to keep one's sanity.

Point 2: It's a misguided public perception that the BLM and cattlemen are aligned. Ranchers are individual types; they go their own way, make their own complaints.

Point 3: Unlike the BLM's many resource problems, the laws and policies for horse management are inadequate. "We've got a few problems with other resources in the state, but nothing like what we have with the horses. Example: The law provides for destruction of the horses, but it's a problem of application. We're just not going to be out there shooting wild horses or we're going to be embroiled in a goddamn public controversy, the damnedest public controversy you can imagine. There's really no way you can destroy a horse in a way that's acceptable to a horse lover."

Point 4: McCormack thinks that any talk about "correct" numbers of wildlife and cows and horses on the range is so much poppycock. "There's just no way you can scientifically come up with meaningful numbers. Any decision we arrive at is political. It's going to be a compromise of some sort."

Point 5: The 1971 wild horse bill wasn't set up to take care of the excess population, only to recognize and maintain what was already out on the range. "The focus of the wild horse people has been to worry about what happens to the horses when we take them off the land. Why don't they worry about what is out there? Once you take them off they're not wild, free-roaming horses anymore, not when you stick them in somebody's two-acre pasture. Those humane societies are destroying millions of dogs and cats that are excessive every year and nobody really thinks much about it. The fact is that there are thousands of domestic horses that are going through the slaughterhouses every year and nobody says anything about that."

McCormack's thoughts on managing mustangs are shared by many in the BLM. "This damn adoption program already takes too much of the money we don't have," one BLMer said to me. "And that's no good with all we've got to do around here. If you ask me, adoptions are the crap end of the horse program. What we should be doing is concentrating our efforts on managing what's out there. We're supposed to be in the business of regulating numbers and keeping them horses healthy, making sure there's enough room for cows and grass and deer and sage grouse. We're even supposed to be worrying about there being enough room for those crazy bastards on their motorcycles. But it's the wrong approach, I tell you. We need to get back to what the federal law says the program is all about, managing a national heritage. *Managing*—not running a horse show for horse lovers and little girls and their fathers."

I asked Tom Owen, the BLM district manager in Carson City, if he agreed with a number of his colleagues around the state that management of the mustangs was the BLM's real job. His reply: "I think all that kind of talk is a lot of bullshit. We're not going to be getting into management. We don't have the money for it. We talked for a long time about managing the herds and selecting for conformation and good color and all that kind of crap. That would cost a fortune."

Point 6: Money. While on the whole, the BLM makes money for the U.S. Treasury, it operates at a $3 million a year deficit in Nevada. The state is timber-poor, it has nothing like the mineral resources of other states, and fees from grazing rights are hardly worth talking about. The horses add to the financial headaches of the Silver State's public lands administrators. This McCormack inevitably gets around to by gingerly noting that Nevada should get a larger share of the annual federal budget to care for horses. The state has well more than half of the nation's mustangs, but it receives much less than half of what Washington allocates to the program.

"How much more do you need to make a difference in what you're doing? To solve this horse problem?"

McCormack buried a hand in his thick grey hair and turned away. There was a long pause. Finally, he said, "We should get a larger share of that pie, but then I recognize you can't arithmetically divide up dollars. It doesn't work that way." He smiled amiably; I sensed that being grey and generous were necessary for self-preservation.

CHAPTER 6

SOLUTIONS

By the mid-1970s, Wild Horse An-
nie was painfully aware that the West's wild horses were multiplying
faster than she had imagined possible. And though she was, by her
own admission, not terribly interested in facts, she was too much a
realist to deny that the horses were magnifying a process of range
deterioration that had begun long before the BLM was born. By the
time the Nevada BLM proposed its first roundup in Stone Cabin
Valley in 1975—which she supported, with the proviso that cattle
also be removed—she was already thinking in terms of "planned
parenthood." She didn't, however, want to use just any method on the
horses. As one of her closest allies would remark several years after
her death: "To Annie you could move the problem, but any kind of
manipulation that destroyed life was anathema. For her, death
wasn't part of the system."

In December of 1976, Wild Horse Annie suggested that vasectomies
be performed on dominant harem stallions. This, she believed, was
preferable to gelding; it would cause no disruption in mustang social
organization and it would allow studs to maintain control of their
harems. Unfortunately, Wild Horse Annie had little knowledge of
biology to back up her belief. For example, she did not know that in
larger harems, more than one stallion often impregnates mares.
Furthermore, there was the problem of identifying and capturing
dominant stallions at a reasonable cost. By 1974, the BLM had
already made a small scale effort at sterilization. The bureau had also

talked about manipulating mares, the factor controlling population growth. It concluded that sterilization or castration would be too difficult, and much too costly.

Helen Reilly has not been deterred by past BLM efforts or the issue of cost and feasibility. One of Wild Horse Annie's first lieutenants, president since 1977 of the International Society for the Protection of Mustangs and Burros (which she says has nineteen thousand members), and a generally well-respected voice on behalf of mustangs, Helen Reilly has frequently come out in favor of "temporary sterilization to maintain sound range management practices."

In the late 1970s, believing that neither castration nor vasectomies was the answer, Helen Reilly promoted the views of Roy Harris and Dave Fly, two veterinarians at the California Polytechnic State University. They suggested that the BLM could practice eugenics by capturing mares and performing a small operation on the least desirable ones: a tiny, flat, capsule inserted under the skin would release a non-toxic hormone, a sort of long-term birth control pill. A second possibility put forth by Harris and Fly involved temporary sterilization of dominant stallions, also by means of a capsule placed under the skin. Since horses would be infertile for only a single breeding season, they would have to be caught every year. Through her organization, Helen Reilly had hoped to get research money for the veterinarians to test their ideas on mustangs. In the end, this specific effort was unsuccessful. Others, however, were working independently along similar lines.

In 1974, John Turner and Jay Kirkpatrick, friends and students in graduate school at Cornell University, were on a backpacking trip in the West. In a Montana saloon, they overheard cowboys complaining about "those damn wild horses." Their interest piqued, and they soon found out what they could about range abuse and the BLM and the management of wild horses. Before long, they began thinking about ways to control mustang numbers. By 1980, when both of them of them had faculty positions—Turner in an Ohio veterinary school and Kirkpatrick at Eastern Montana College—they were able to convince the Department of Interior to give them $200,000 to test an antifertility drug on mustangs. Initially, they tried out four different drugs on twenty-four ponies in an equine research lab at the University of Pennsylvania. On the basis of this work, they settled on a steroid hormone which, they believed, had several characteristics to recommend it: it induced infertility by severely reducing sperm production; a single injection was effective for a whole breeding season; the infertility was reversible; and the drug did not interfere with a stallion's basic sex drive.

In the mid-1970s, Turner and Kirkpatrick had collected data on the reproductive behavior of wild horses in Montana's Pryor Mountains, and it was here that they had hoped to test their drug. But just as they were about to move ahead, their effort was blocked by wild horse protectionists. Turner says, "Some horse-protection organizations spread misinformation about our project, arousing such strong hostility among the local people that we had to leave." In 1980, they moved their work to Challis, Idaho. There they tried their drug on ten stallions just prior to the breeding season.

The procedure was simple. The steroid hormone was injected into the horses with small capsules shot by a dart gun from a helicopter. After that, the temporarily sterilized stallions and their bands were closely monitored. On the basis of these observations it was concluded that the antifertility drug worked. In the study bands, there was an eighty percent decrease in foal production, and sperm counts rebounded after the drug wore off. That showed that infertility was temporary. The new foal crops looked healthy, and there was no apparent aberrant behavior in bands led by a temporarily infertile stud.

Turner claims that the antifertility procedure would be cheap. Reporting results in the February 1984 issue of *Smithsonian* magazine, he says, "Two persons using a helicopter and a dart rifle could drug a five hundred horse herd for an entire year in just two days, and cut population growth in half. The cost and stress on the animals would be a mere fraction of that required for adoption."

Turner and Kirkpatrick's work has not been received enthusiastically by biologists familar with wild horses. Two principal objections have been raised. The assumption that a single dominant stud does all of the breeding in a wild horse band isn't always true, particularly in large bands of eight or more horses where two or three males may father offspring. Also, Turner and Kirkpatrick proceeded on the assumption that mares are faithful to a single band. But other scientists who have watched mustangs in Wyoming, Oregon, and Montana have seen mares move among harems. Still others have complained that this birth control method blocks gene flow from the genetically superior animals.

In 1982, the National Research Council brought together fourteen biologists to evaluate research on wild horses. They concluded that to carry out the kind of program advocated by Turner and Kirkpatrick "would appear to be prohibitively costly in terms of time, manpower, and flight costs." The committee of scientists added, "On the whole, there appear to be many problems and uncertainties with the method as a general technique for West-wide population control."

The BLM largely agrees with these conclusions, adding that supressing fertility in mares appears to be more promising. In a briefing report directed at Turner's *Smithsonian* article, the BLM says, "The advantages of mare-focused population control is that treatment of a given percentage of mares in a herd could be expected to result in an equal reduction in reproductive success of the herd and it does not block gene flow from genetically superior animals."

From Turner's perspective, "The issue of wild-horse management is a political one. Many BLM people have worked hard to make the Adopt-A-Horse program an ongoing concern, and some of them have no enthusiasm for the drug approach."

Turner is right, I believe, in noting that management of the mustangs is a political matter, but wrong in laying blame on those who administer the Adopt-A-Horse program.

Dawn Lappin had been on a horse only two or three times in her life, and her only experience in caring for animals before going to work for Wild Horse Annie in the 1960s was with what she calls household pets: dogs, snakes, turtles. She began helping Annie by typing addresses on envelopes, and then moved on to play a central role in developing Nevada's, and the nation's, wild horse adoption program. But for all her involvement, as late as 1977 when Wild Horse Annie died, Dawn admits that she didn't even know how the BLM allocated range forage to livestock and wildlife. "In those days we were making our arguments on a purely emotional basis." But all this changed quickly when Dawn found herself trying to step into Annie's gigantic shoes as president of WHOA—Wild Horse Organized Assistance. She started reading everything she could get her hands on, and before long she was doing battle with ranchers, the BLM, and other activists who had a different vision for the horses.

For a couple of years after Wild Horse Annie's death, Dawn carried out her advocacy role in an office in downtown Reno. Then in 1979, when the expense became unmanageable, she moved into a basement studio in Annie's home, occupied by Annie's eighty-nine year old mother. Cluttered with BLM reports, environmental impact statements, and piles of correspondence and field notes, her office is a reminder of her abiding affection for Wild Horse Annie. Above her desk is a Nevada license plate stamped WHOA–1. On a nearby piano is an engaging bronze of Wild Horse Annie alongside a mustang. Not far away is a wall covered with reminders of the diminutive, polio-stricken woman who symbolized the wild horse movement: photographs, silver and bronze plaques from clubs and chambers of commerce, autographed copies of legislation she engineered, and letters of commendation from U.S. presidents, senators, and congressmen.

Short, stocky, round-faced with dark eyes and a ruddy complexion, a chain-smoker with a deep voice, Dawn Lappin today is larger than WHOA itself. With due respect to the ten thousand or so people who provide her with financial support, without Dawn's indefatigable efforts on behalf of the horses, WHOA would be much less effective. She has made it her business to know scores of Nevada ranchers on a first-name basis. She has hiked their ranches, she has counted cows and horses, she has taken note of horse trails and the size of stud piles, she has done her best to assess range damage. Some ranchers listen to her, and some will even sit with her and try to find a middle ground on seemingly irreconcilable issues. Dawn Lappin has also cultivated several informants around the state who keep her up to date on when and where the next illegal mustanging will occur, and when the BLM is doing something "suspicious," something that ought to be looked into right away.

Dawn Lappin has "Deep Throats" within the BLM who constantly feed her information on proposed roundups and policies that affect mustangs, and she sometimes even gets access to bureau decisions before district managers have heard about them or have had a chance to carry them out. If unhappy with an impending policy decision, Dawn can easily get the ear of the Nevada's BLM associate director, Roger McCormack, or the state's director, Ed Spang. She will not only be heard, but her stentorian threats have gotten roundups stopped, decisions delayed, policies altered, even changed. Just about anyone worth talking about in the Nevada BLM who has to deal with wild horses respects her knowledge, her concern, her willingness to learn and change. Wild horse specialists and many within the BLM sometimes don't agree with Dawn Lappin, and some don't care for her bluntness and critical tongue. But few ignore her.

Dawn Lappin has a pretty good understanding of the BLM and her own image. One minute she says she doesn't trust the bureau and is not above blackmailing bureau chiefs with letters of protest from WHOA's members to get what she wants. In another moment, she says she has a "love-hate relationship" with the agency. "I respect and admire a great many people in the BLM, but some of those people are into playing games—that's what I call it. They tell you one thing and five minutes later they're doing something else."

Not afraid to admit that she has made mistakes, nor fearful of a confessional revelation, she says, "The BLM and a lot of other people see me as a flake, a nut, a goody two-shoes. That's because I understand the need for big business, because I understand the need for mining, because as a westerner I have certain attachments to the ranching industry. I've been part of its history too. I was born in Reno, but my family comes from Ely. I come from three generations of Nevadans. One side of my family was into mining, the other side was into ranching. It's a heritage I wouldn't want to see lost. I concentrate my efforts on Nevada because it's the most controversial spot. Nevada has good and caring people, and I don't like people seeing it just in terms of prostitution and gambling and bad treatment of horses."

"What do you want in the long-run for the horses?" I asked her.

"I want a better attitude. And I want the BLM to grasp the idea that it's not dealing with a horse problem or a cow problem. It's a *grazing* problem out there."

"What others issues do you see as critical to proper management of the West's wild horses?"

She identified three: "the lousy political climate—everything in the BLM is politics; the lousy belief held by many ranchers that access to public land is a right rather than a privilege; and the impact of *all* grazing on our fragile ecosystem."

I once asked her how she saw her role. "If you can summarize it?"

"Idealism is what I fight. Without WHOA there would only be extremes. Only roundups or no roundups. I don't like anything starving to death. I would rather risk losing a portion than everything. Maybe my attitude comes from living in a gambling state."

Late one afternoon, after she had graciously spent several hours going back over old questions and talking candidly about everything and just about everyone remotely connected with the wild horse program, I asked her if she was in favor of some form of birth control for the mustangs.

"We've been trying to get them to do the same thing they've been doing for years with wildlife and that's manipulating the population." She added that she doesn't want horses interfering with wildlife, and she doesn't want them "bankrupting the livestock industry."

"Would you go along with a plan that would reduce Nevada's wild horse population to what it was in 1971 when the federal government took control of the horses?"

One answer I got was that she hadn't come up with a number, adding that it was the BLM's job to first determine the nature of the available resources and then to calculate the proper number of horses to keep the range healthy. On another occasion—perhaps I caught her in a different mood—she said that she would agree to a reduction of the state's wild horse population to seven thousand. But before I could pursue the question of how she arrived at a figure lower than the number in the state in 1971, she threw out her bottom line. "However, if I agreed, you would not see the BLM reduce the number of cows. And I would only agree if the number of cows were reduced."

I'd been hiking in the Desatoya and Clan Alpine mountains for more than a week, and all the while I felt privileged, exultant, riding a high that daily seemed to be rising and swelling. There was hardly a moment when one of nature's creations didn't come into view or spring forth from recent memory: fluttering sage grouse and chukar by the bushel, short-tailed pinyon jays and whisking chickadees, falcons that circled and drifted and dived for prey, many-pronged deer that stood tall and proud, once a bobcat slinking into cavernous rocks. Now and then even a prowling coyote, as suspicious of me as I was of it. Here and there, I'd see a cinquefoil or a hollyhock, a skullcap, other wild flowers I couldn't name. During this devouring nonce, I saw nary a human. I did see quite a number of languorous cows. And plenty of wild horses.

Then one morning, camped on the narrow bench of a steep mountain, I woke with first light and made a fire and put on water for my caffeine addiction. After the brown juice pumped up my heart and brought my brain to life, I sat lotus-like on a rock to survey my surroundings. On distant slopes in three directions and in a boxy canyon bottom I saw a familiar sight: several dozen mustangs enjoying breakfast. I watched them for ten or fifteen minutes, recalling what they ate in these mountains, how their appetites compared with mine, with other animals. I thought of what I had learned from the BLM, from ranchers, from conservationists, from wild horse activists, from my own observations and reading. And then, as if it were finally time to bring a modicum of seriousness to my indulgent, carefree tramping, I recalled a now-classic article written by the esteemed biologist, Garrett Hardin, that I first read while a graduate student at Syracuse University.

In the article, entitled, "The Tragedy of the Commons," Hardin opens with a quote from two scientists who note that, in their "considered professional judgment," the problem of nuclear war has no technical solution. A technical solution, Hardin says, is "one that requires a change only in the techniques of the natural sciences, demanding little or nothing in the way of change in human values or ideas of morality." Hardin says that there is a class of "no technical solution problems," among which he includes the "population problem—as conventionally conceived." The difficulty is that the world is finite, and a finite world can only support a finite population. At some point, population growth must equal zero. Hardin then goes on to point out that when this condition of zero population growth is finally met, Bentham's goal of "the greatest good for the greatest number" cannot be realized. One reason, a mathematical one sufficient unto itself, is that it is impossible to maximize two or more variables at the same time.

Another reason we cannot have "the greatest good for the greatest number" is a matter of the laws of biology and physics. If humans were to maximize their numbers by putting all available energy into simple maintenance of the species, there would be no calories left for even the most simple amenities of life. Man would be precariously perched on a wafer between mere existence and unremitting primal need. As Hardin concludes, "Maximizing population does not maximize goods. Bentham's goal is impossible." The ideal population should not be defined in terms of a maximum, but rather an optimum—however difficult it may be to arrive at that optimum.

Garrett Hardin notes that Westerners down to the present have been afflicted by Adam Smith's notion of the "invisible hand," the idea that an individual in pursuit of his own best interest is led by an invisible hand which promotes the public interest. But, Hardin asks, does a laissez-faire policy in reproduction lead to an optimum population? And if not, which of our freedoms is genuinely defensible?

Hardin then sketches a scenario first put forth by an amateur mathematician named William Forster Lloyd in 1833. The scenario Hardin calls "the tragedy of the commons," in which tragedy refers to—in the words of the philosopher Alfred North Whitehead—"the solemnity of the remorseless working of things." At this point, I can do no better than let Hardin describe how the tragedy of the commons develops.

"Picture a pasture open to all. It is to be expected that each herdsman will try to keep as many cattle as possible on the commons. Such an arrangement may work reasonably satisfactorily for centuries because tribal wars, poaching, and disease keep the numbers of both man and beast well below the carrying capacity of the land. Finally, however, comes the day of reckoning, that is, the day when the long-desired goal of social stability becomes a reality. At this point, the inherent logic of the commons remorselessly generates tragedy.

"As a rational being, each herdsman seeks to maximize his gain. Explicitly or implicitly, more or less consciously, he asks, 'What is the utility to *me* of adding one more animal to my herd?' This utility has one negative and one positive component.

"The positive component is a function of the increment of one animal. Since the herdsman receives all the proceeds from the sale of the additional animal, the positive utility is nearly $+1$. The negative component is a function of the additional overgrazing created by one more animal. Since, however, the effects of overgrazing are shared by all the herdsmen, the negative utility for any particular decision-making herdsman is only a fraction of -1.

"Adding together the component partial utilities, the rational herdsman concludes that the only sensible course for him to pursue is to add another animal to his herd. And another; and another. . . . But this is the conclusion reached by each and every rational herdsman sharing a commons. Therein is the tragedy. Each man is locked into a system that compels him to increase his herd without limit—in a world that is limited. Ruin is the destination toward which all men rush, each pursuing his own interest in a society that believes in the freedom of the commons. Freedom in a commons brings ruin to all."

Does this scenario bring into sharper focus human dealings with the West's wild horses? And does it turn us in the direction of equitable and sensible solutions?

Ranchers will, when pressed hard enough and often enough, give; they will allow some reduction in the number of cows they are permitted to run on the public domain. But they will not give much, and I strongly suspect that, like the self-serving herdsmen of Hardin's scenario, they will give greater consideration to the proceeds from the sale of a single animal than they will to the incremental but inevitable successional changes and desertification that would eventually lead to the ruin of all.

Ranchers have their counterparts in ardent horse lovers and many animal protectionists who do not think or cannot find the time to see what is at stake. Some would have the West's wild horse herds grow until their numbers reach a biological maximum. Or beyond. At which point, the ranchers are gone and the livestock is gone and the wildlife is gone and the desert has turned to rank weeds and withering, starving horses. Few, of course, contemplate this conclusion to their actions and choleric letters to congressmen; that is the problem.

Then there is the inescapable fact that grasses and soils are unsexy. They are difficult to know, and even more difficult to appreciate.

We can barely conceive of how to express attachment and affection toward them. There are no best sellers or prize-winning children's books on desert soils and Great Basin grasses; the horses of child-lore gallop on an undifferentiated plain, and they don't eat. Few people seem to know that soils, once lost, may take hundreds of years to recover, and some can never be brought back to life. Few know that a "natural" ecosystem—one that is diverse and stable and relatively pristine, different and beautiful and untrammeled as it was before man and his domesticated and introduced animals came on the scene—is equally difficult to make whole. And so what?, many ask. After all, Judeo-Christian teachings have given man dominion over the earth and all its creatures.

Still others, among them some ostensibly well-informed conservationists, dwell in a world that can only be described as one filled with mystic light: a problem with horses, they say, is really a problem with cows and we will tolerate no solution to the horse problem unless there is simultaneously a solution to the too-many-cows problem. The problem is that this kind of activist is so firmly determined to solve a complex issue in one bold stroke that he, or she, would rather gaze in horror from campsite while gene-driven horses and profit-driven cattlemen ineluctably move toward a predictable denouement than step outside the gauzy bubble. Every day, every week, every month that goes by without meaningful dialogue among zealous activists, hardline ranchers, and the BLM is one more setback for the very things these kinds of conservationists most desire: nature noble and abundant.

Contrary to prejudices I'd acquired from social scientists, I met many in the BLM who were knowledgeable about ranching and grasses and wild horses, and they also impressed me as genuinely concerned with proper management of our public lands. Unfortunately, most of these same individuals are sweaty laborers: they talk to ranchers, they ride the range, they measure and assess damage, they round up horses, they occasionally get kicked in the teeth. Some of them are, I think it can fairly be said, public servants in the best sense of the word. Or, rather, as good as one can hope for in a decidedly imperfect world.

But these sun-burned cowboys have no clout to speak of: their opinions and their field reports have as much heft as clubs made of ostrich plumes. I've received the distinct impression that as soon as one of them makes it into the ethereal policy- and decision-making corridors of the bureau, he suddenly begins to lose his sense of priorities. He acquires a gift for gab, political savvy, and an exasperating sense of procrastination. Nothing gets done, or it gets done too late. Meanwhile, those with the strongest voices and the best lawyers and the largest constituencies go about demanding and attacking the BLM, veiling self-interest in the name of "the public's interest." Pushing for horses, or for cows, or for wildlife. Throwing out vulgar prejudices as truth or reasoned argument.

The horses, of course, are victims. They have no moral order other than that embodied in the forces of nature and natural selection: those stallions and those mares which produce the most offspring reign supreme in nature's scheme of things. In a sense, they rule the mountaintops and the valley bottoms because in past generations, their ancestors' genes were disproportionately represented in the population. The inevitability of natural selection is as old as all things living and dead. It is a law at once supremely selfish, relentless, and ultimately uncaring. The logic of natural selection is brutally simple, its forces and dictates remorseless. And never more so than when nature has lost its checks and balances—predators that prevent a dominant species from destroying both its environment and itself. Man and mustang have a good deal in common.

No, I don't think that the present problem with wild horses is hopeless, or that those who are involved with them have no intuitive sense of the tragedy of the commons. But with each passing hour in which too many mouths feed on fragile grasses and drink scarce water, the chances of reversing man's stupidity decreases. Since man has drastically changed an ecosystem by introducing animals without natural predators, he must now rationally act as the predator he is.

I have already indicated where my some of my priorities lie. At the risk of repeating myself, it may be helpful to line them up and clearly flag my guiding criteria.

Little-disturbed soils and native grasses are among the most precious things we have left in the American West. They deserve our primary attention, even ahead of wildlife—if such a choice need be made—because drastically altered they are the most difficult to revive. And, in any event, their period of convalescence is long indeed. Without good soil, plant life is meager or nonexistent. And without plant life there is no wildlife, no horses, no cows, no humans. We do not have to love that which we place first on a list of priorities; we only need understand the necessity of the choice.

I am inclined to put wildlife in its many forms next on my list. These earthlings have every bit as much right to live and procreate and run free in spacious habitats as does man. More, one might argue—given man's insistent proclivity to destroy the earth for all living things. I stand with many conservationists who, when faced with a choice between wildlife and domesticated species, come down on the side of deer and antelope and bighorn sheep. Again, it seems to me, the issue has a good deal to do with the ease or difficulty of recovering losses. And, it might be added, whether the creatures are or were part of the natural order rather than a product of man's needs and genetic tinkering. I do not denigrate the value of a horse's life, and I wish I didn't have to compare the value of lives. Faced with the dilemma of one in preference to another, I choose those most genuinely wild and first on the scene. The crude reality is that were all of the nation's wild horses eliminated tomorrow, the Great Basin could easily be restocked with mustangs in today's numbers in less than half a dozen years.

Ranchers come next on my list. But not any rancher, and not every rancher. Viable desert ranching requires vast open spaces—hundreds of thousands of acres—and, notwithstanding the mindless, abusive ways of some, it is within such enormous expanses that native soils and native grasses and wildlife have the best chance of being preserved as other than zoo-like relics. Allow citified types to discover and thoroughly colonize and desalinize the West's deserts, and before long, the only thing native in them will be the offspring of the first to establish their suburban ranchettes. Ranching as a way of life also deserves our vote because it represents the frontier, a weighty and romantic part of our history, a symbol of perseverance and individuality and free-thinking, a reminder of wagon-wheel democracy on the roll. Ranching is part of the wholesome myth that allows us to see ourselves as young, vibrant, hopeful, growing, and at once crassly self-absorbed and yet open to redemption.

Having said this, it is no less true that ranching the public domain rankles and plays havoc with our sense of trespass, and not without reason. Some ranchers, like some people, have no land or wildlife ethic. They are Hardin's herdsmen. They unflinchingly take all that they can get. But they can be checked, and matters can be improved in ways that benefit the natural order.

Since the 1934 Taylor Grazing Act, the Grazing Service and its successor, the BLM, have had a mandate from the people to properly care for public lands. The BLM has not done a commendable job. With proper leadership, the BLM can conduct range surveys that will stand up to scientific scrutiny. It can reduce the number of cows and sheep that use our lands, it has the power to eliminate abusive ranchers, and it could, if it wanted to, give thoughtful ranchers larger public allotments. Large enough so that they could live comfortably and have no urgent need to take advantage of soils and plants and wildlife.

We have a problem with too many cows on our public lands, and many of them ought to be removed. At the very least, taken off in sufficient numbers to maintain the stability of our frail desert ecosystems. Even better, brought down to levels that would allow for the reintroduction and gradual recovery of plants and animals more nearly representative of the desert before the appearance of rapacious man. All of this is an issue now decades old, in fact, one that long predates the idea of a federal agency to watchdog the nation's common. What should not be done, I believe, is to confuse the historical failings of an emasculated and misunderstood agency with the growing problem of too many wild horses. Let's first handle the more tractable problem of horses; there is extant, explicit legislation that permits solutions, and there is proposed federal legislation that will provide further solutions to affrayments over what to do with excess or unadoptable horses. Brewing compassion for mustangs with concern for salty and sandy soils only begets a debilitating broth.

Today, some in the BLM believe that the agency missed a favorable opportunity to reduce the West's cattle and sheep population in 1971. At that time, the bureau could have explicitly reserved forage for wild horses by reducing the number of livestock. They didn't do so then or in the years to come, because the bureau "has been obsessed with getting sale authority for horses that it couldn't adopt."

The BLM frequently gives lip service to its job of "ensuring viable populations of healthy free-roaming wild horses," and the maintenance "in good ecological condition of the basic soil, water and vegetation resources that comprise the rangeland ecosystem." But the bureau can be immoderately slow-going. In 1974, the BLM lost a lawsuit to the National Resources Defense Council, which required the BLM to prepare environmental impact statements on western rangelands. These statements were to include both the grazing impact of cows and also of wild horses. By 1984, the Nevada BLM had come up with plans for less than ten areas for managing wild horse herds: ten out of a total of one hundred thirty-nine. Currently, the Nevada BLM is only producing about half a dozen environmental impact statements a year. The California BLM is moving at a somewhat quicker pace, but not fast enough to win any awards. By the middle of 1984, it had worked out optimal wild horse herd figures for less than twenty percent of its forty herd management regions.

Many areas in the West where mustangs live are unfenced. If the propagation of wild horse herds is not regulated better than it has been thus far, it is almost certain that ranchers will continue to bring lawsuits against the BLM for trespass and erosion of their public grazing allotments. Based on precedent, it is fair to assume that the ranchers will win the majority of these suits. If judges decide, as in the case won by the Fallinis, that mustangs have no right whatsoever on private property, or, as with the Sweetwater Ranch, that water is no different than real property, then the BLM could be faced with a decision either to eliminate wild horses from many areas or consider fencing them in. Fencing would almost certainly affect the natural movements of horses and wildlife, possibly have dire effects on vegetation, and most certainly dilute the symbolic value we place on open space.

In the last couple of years, the BLM has recognized the need to reduce the wild horse population to a figure close to the number on the range in 1971 when the federal law protecting the horses was passed. Then, and for roughly a decade thereafter, that figure was thought to be seventeen thousand. Willing to learn from professional biologists, and willing to admit that it was only guessing how many mustangs were in the West when given its unprecedented task, some within the bureau now say that the 1971 number was too low by at least five thousand. Today, the BLM apparently sees no problem in accepting this higher figure for the number of mustangs in the West in 1971, and in a few years it may once again alter its thinking— upward. By late 1984, some BLMers in Nevada were having to confront alarming numbers. In the mountains of the Carson City BLM district, for example, helicopter-borne cowboys counted fifty-two hundred mustangs in 1983. When a more careful and more sophisticated census was taken a year later, the figure had jumped to seventy-eight hundred, an increase far too great to be accounted for by simple reproduction. According to one knowledgeable wild horse expert in Carson City, even seventy-eight hundred is probably short of the real number by at least one thousand horses. Based on this kind of information, the biology of horse reproduction, and the word of professional biologists who study growth rates in animal populations, I think that the 1971 base figure may have been too low by ten thousand. Quite possibly more. Horses simply cannot have offspring as fast as indicated by a combination of the number still roaming free and the figure of those put up for adoption.

Unhappily, the BLM's wild horse program has started to resemble a sinking ship with too many passengers on board. The population of wild horses in the West in 1984 may well exceed seventy thousand, there are no signs whatsoever that dominant harem stallions are at all concerned about planned parenthood, and mustangs taken in roundups are increasingly difficult to find homes for. No one in the BLM, to my knowledge, is seriously considering using that proviso in the federal law which allows the bureau to kill the mustangs when homes cannot be found for them. Putting mustangs to sleep with a painless injection is one course of action that one and all, including the national director of the BLM and the Secretary of Interior, agree would cause a national furor of unmanageable proportions. By mid-1984, many bureau managers, and plenty below them, were starting to talk about "maintenance absorption" of the bureau's wild horse budget. It costs more than two dollars a day to board a wild horse, and when several thousand of them stay around for several months it doesn't take long before other budget items in the wild horse program are affected: for roundups, for completing herd area studies, and for promoting the adoption program and setting up new adoption centers.

Despite new adoption centers in the East and the use of temporary facilities in cities with untapped demand for mustangs, it is becoming more and more difficult to find adopters. The adoption fee—one hundred twenty-five dollars—is not so high that people are discouraged from taking a horse, at least not those who will properly care for one. Many of the horses are just too old or not very attractive, and they can only be adopted out at considerable promotional expense. Some activists argue that the BLM is not trying hard enough to find homes for the horses, but these critics seem to have little understanding of how expensive it is to advertise anything.

Most in the bureau feel that the only reasonable long-run solution is federal legislation that would allow for the open-market sale of unadoptable and excess horses. I agree. I agree, knowing that most of the mustangs sold in this manner will wind up as pet food or on foreign plates. With some reluctance, I am forced to agree with Garrett Hardin and others who, at this point, would invoke a principle of situational ethics: the morality of an act is a function of the state of the system at the time it is performed. Once we have allowed ourselves, or have allowed those who manage for us, to create a crisis-threatening situation, one destructive not just to horses but to other living things, we have to take corrective measures even if they're distasteful. Action delayed only magnifies stern necessity.

No one of good conscience wants to stand by while horses or other animals starve to death: that is cruel. If we are willing to lend an ear to Montaigne, that ever-contemporary, sixteenth-century French humanist, we learn that cruelty is far worse than the seven deadly sins. It is the *summum malum*, it is as unbearable as the sight of the instruments of torture in Giotto's *Last Judgment.* Montaigne thought that the ultimate victims of human cruelty were animals. He abhored the way men lord over them. In a fit of misanthropy and anger, he wrote that little could be more absurd than "this little miserable and puny creature, who is not so much as master of himself . . . should call himself master and emperor of the universe."

In 1984, Congress was unable to pass an amendment to the Wild Horse and Burro Act that would allow the BLM to sell excess horses. But at the last minute, on October 10, it came up with an unusual compromise that may eventually have the same result. In addition to appropriating five million dollars for the next fiscal year to run the wild horse program, Congress also put up eleven million dollars for the immediate gathering of seventeen thousand mustangs, ten thousand of them in Nevada. The bill notes that eleven thousand of the total are "excessive wild horses" that should have been rounded up in 1983 and 1984. The remainder are also excessive, horses the BLM didn't remove during the two prior years because of budget constraints and the lack of adopters. When these horses are captured they will be boarded in privately owned feed lots. It would be surprising if more than a small fraction of them find homes. Senator James McClure from Idaho and other congressmen who have been sponsoring new federal legislation that would allow for the sale of mustangs by the BLM are betting that when taxpayers realize that they're paying up to a million dollars a month to feed and house mustangs, they may finally decide that this is too high a price to pay for a cultural prejudice.

Slowly I came down from the high ridge, followed a well-used horse trail part way around an eroded hump shaved like an army recruit, then turned into a narrow, rock-strewn creek. Suddenly, hearing a rustle in squat juniper behind me, I turned, glanced upward, and there on the slope I'd just come down was a herd of fluffy-tailed antelope dashing for high ground and cover. There were eight, ten, fifteen—I don't how many, it happened that fast—and they were flying. Delicate legs in harmony, gone before I could gather my senses and fully hoard the image. For a long moment, I sat on my pack and savored what I'd seen. I turned back time and put the herd back into the branching, many-angled space. I licked the inviting smell of low sage, wondered where they'd spend the night. It was my last evening in the Clan Alpines; I admonished myself for not discovering unknown Nevada at a younger age.

The sky now full of angelic purple light, I hurried down the creek bed. I crossed a familiar trail, stepped over numerous stud piles, and then wistfully peered up toward the summit where I'd watched horses drink and play several nights before. I walked into a welt-high pool and followed a ribbon of water between rock walls. Then, as I came into the open, he came into view. He stood off to my right not fifty yards away. Silent, head high, facing me head-on. A chill coursed up my back, and I wondered if he'd been waiting for me. He wasn't very big, but he looked mighty big in his aloneness on that flat hill that night. I stood and stared, went a step or two forward, took another glance, moved on, took still another, and all the while he didn't move. I couldn't be sure, I'll never be sure, but I'd swear he was the same bachelor that I'd seen on a roller coaster hill on another evening as I began to make camp. Then he had also stared and taken me in—for a short, long time. But now, knowing I had to hurry on before losing the last light and losing my way, I scampered down the ravine toward my pickup. I only looked back once. He was still there; he was still motionless. I squinted and I prodded the intellectual side of my brain. No matter what, I wanted him to be able to stay there.